T0258616

Handbook of Doppler Radar Observations

Handbook of Doppler Radar Observations

Edited by **Henry Collier**

New York

Published by Callisto Reference,
106 Park Avenue, Suite 200,
New York, NY 10016, USA
www.callistoreference.com

Handbook of Doppler Radar Observations
Edited by Henry Collier

International Standard Book Number: 978-1-63239-384-5 (Hardback)

Printed in the United States of America.

Contents

Preface

Doppler radar systems have been extremely helpful in enhancing our perception and monitoring capabilities of atmospheric phenomenon. Scientists and practitioners are using its methods and radio detecting and finding techniques, like wind profilers and others, frequently for analysis and operations. Some of the topics discussed in this book are – extreme weather supervision, wind and turbulence retrievals, estimation of precipitation and nowcasting, and the volcanological applications of Doppler radar. This book is appropriate for graduate students who are seeking an initiation in the area or for experts wanting to refresh their knowledge and information on the same.

This book unites the global concepts and researches in an organized manner for a comprehensive understanding of the subject. It is a ripe text for all researchers, students, scientists or anyone else who is interested in acquiring a better knowledge of this dynamic field.

I extend my sincere thanks to the contributors for such eloquent research chapters. Finally, I thank my family for being a source of support and help.

Editor

Part 1

Doppler Radar and Weather Surveillance

Doppler Radar for USA Weather Surveillance

Dusan S. Zrnic

NOAA, National Severe Storms Laboratory
USA

1. Introduction

Weather radar had its beginnings at the end of Word War II when it was noticed that storms clutter radar displays meant to reveal enemy aircraft. Thus radar meteorology was born. Until the sixties only the return power from weather tracers was measured which offered the first glimpses into precipitation structure hidden inside clouds. Possibilities opened up to recognize hail storms, regions of tornadoes (i.e., hook echoes), the melting zone in stratiform precipitation, and even determine precipitation rates at the ground, albeit with considerable uncertainty.

Technology innovations and discoveries made in government laboratories and universities were quickly adopted by the National Weather Service (NWS). Thus in 1957 the Miami Hurricane Forecast Center commissioned the first modern weather radar (WSR-57) the type subsequently installed across the continental United States. The radar operated in the 10 cm band of wavelengths and had beamwidth of about 2°. In 1974 more radars were added: the WSR-74S operating in the band of 10 cm wavelengths and WSR-74C in the 5 cm band.

Development of Doppler radars followed, providing impressive experience to remotely observe internal motions in convective storms and infer precipitation amounts. Thus scientists quickly discovered tell tale signatures of kinematic phenomena (rotation, storm outflows, divergence) in the fields of radial velocities.

After demonstrable successes with this technology the NWS commissioned a network of Doppler radars (WSR-88D=Weather Surveillance Radars, year 1988, Doppler), the last of which was installed in 1997. Much had happened since that time and the current status pertinent to Doppler measurements and future trends are discussed herein.

The nineties saw an accelerated development of information technology so much so that, upon installation of the last radar, computing and signal processing capabilities available to the public were about an order of magnitude superior to the ones on the radar. And scientific advancements were still coming in strong implying great improvements for operations if an upgrade in processing power were to be made. This is precisely what the NWS did by continuing infusion of the new technology into the system. Two significant upgrades have been made. The first involved replacement of the computer with distributed workstations (on the Ethernet in about 2002) for executing algorithms for precipitation estimation, tornado detection, storm tracking, and other. The second upgrade (in 2005) brought in fully programmable signal processor and replaced the analogue receiver with the digital receiver. In 2009 the NWS started the process of converting the radars to dual polarization which should be accomplished by mid 2013.

The number of radars used continuously for operations is 159 and there are two additional radars for other use. One is for supporting changes in the network brought by infusion of new science or caused by deficiencies in existing components (designated KCRI in Norman, OK). The evolution involves both hardware and software and the update in the former are typically made annually. The other (designated KOUN in Norman, OK, USA) is for research and development. Therefore its configuration is more flexible allowing experimental changes in both hardware and software.

Conference articles and presentation about the WSR-88D and its data abound and there are few descriptions of its basic hardware. Very recent improvements are summarized by Saxion & Ice (2011) and a look into the future is presented in Ice & Saxion (2011). Yet only few journal articles describing the system have been published. The one by Heiss et al. (1990) presents hardware details from the manufacturer's point of view. The paper by Crum et al. (1993) describes data and archiving and the one by Crum & Alberty (1993) contain valuable information about algorithms. The whole No. 2 issue of Weather and Forecasting (1998), Vol. 13 is devoted to applications of the WSR-88D with a good part discussing products that use Doppler information. A look at the network with the view into the future is summarized by Serafin & Wilson (2000).

As twenty years since deployment of the last WSR-88D is approaching there are concerns about future upgrades and replacements. High on the list is the Multifunction Phased Array Radar (MPAR). At its core is a phased array antenna wherein beam position and shape are electronically controlled allowing rapid and adaptable scans. Thus, observations of weather (Zrnic et al., 2007) and tracking/detecting aircraft for traffic management and security purposes is proposed (Weber et al., 2007). Another futuristic concept is exemplified in proposed networks for Cooperative Adaptive Sensing of the Atmosphere (CASA) consisting of low power 3 cm wavelength phased array radars (McLaughlin et al., 2009).

Very few books on weather radar have been written and most include Doppler measurements. Here I list some published within the last 20 years. The one by Doviak & Zrnic (2006) primarily concentrates on Doppler aspects and contains information about the WSR-88D. The book by Bringi & Chandrasekar (2001) emphasizes polarization diversity and has sections relevant to Doppler. Role of Doppler radar in aviation weather surveillance is emphasized in the book by Mahapatra (1999). The compendium of chapters written by specialists and edited by Meishner (2004) concentrates on precipitation measurements but has chapters on Doppler principles as well as application to severe weather detection. Radar for meteorologists (Rinehart, 2010) is equally suited for engineers, technicians, and students who will enjoy its easy writing style and informative content.

2. Basic radar

The surveillance range, time, and volumetric coverage are routed in practical considerations of basic radar capabilities and the size and lifetimes of meteorological phenomena the radar is supposed to observe. This is considered next.

2.1 Considerations and requirements for storm surveillance

Table 1 lists the radar parameters with which the surveillance mission is supported. Discussions of the reasons behind choices in volume coverage and other radar attributes of the WSR-88D network, with principal emphasis on Doppler measurements, follows.

Requirement	Values
Surveillance: Range Time Volumetric coverage	 460 km < 5 min hemispherical
SNR	> 0 dB, for Z= - 8 dBZ at r=50 km (exceeded by ~5 dB)
Angular resolution	$\leq 1^{\circ}$
Range sampling interval: For reflectivity For velocity	 $\Delta r \leq 1$ km; $0 < r \leq 230$ km; $\Delta r \leq 2$ km; $r \leq 460$ km $\Delta r = 250$ m
Estimate accuracy: Reflectivity Velocity Spectrum width	 ≤ 1 dB; SNR>10 dB; $\sigma_v = 4$ m s^{-1} ≤ 1 m s^{-1}; SNR> 8 dB; $\sigma_v = 4$ m s^{-1} ≤ 1 m s^{-1}; SNR>10 dB; $\sigma_v = 4$ m s^{-1}

Table 1. Requirements for weather radar observations.

2.1.1 Range

Surveillance range is limited to about 460 km because storms beyond this range are usually below the horizon. Without beam blockage, the horizon's altitude at 460 km is 12.5 km; thus only the tops of strong convective storms are intercepted. Quantitative measurements of precipitation are required for storms at ranges less than 230 km. Nevertheless, in the region beyond 230 km, storm cells can be identified and their tracks established. Even at the range of about 230 km, the lowest altitude that the radar can observe under normal propagation conditions is about 3 km. Extrapolation of rainfall measurements from this height to the ground is subject to large errors, especially if the beam is above the melting layer and is detecting scatter from snow or melting ice particles.

2.1.2 Time

Surveillance time is determined by the time of growth of hazardous phenomena as well as the need for timely warnings. Five minutes for a repeat time is sufficient for detecting and confirming features with lifetime of about 15 min or more. Typical mesocyclone life time is 90 minutes (Burgess et al., 1982). Ordinary storms last tens of minutes but microbursts from these storms can produce dangerous shear in but a few minutes. Similarly tornadoes can rapidly develop from mesocyclones. For such fast evolving hazards a revisit time of less than a minute is desirable but not achievable if the whole three dimensional volume has to be covered. The principal driver to decrease the surveillance time is prompt detection of the tornadoes so that timely warning of their presence can be issued. Presently, the lead time for tornado warnings (i.e., the time that a warning is issued to the time the tornado does damage) is about 12 minutes (see Section 5).

2.1.3 Volumetric coverage

The volume scan patterns currently available on the WSR-88D have maximum elevations up to 20° and many are accomplished in about 5 minutes. Meteorologists have expressed a desire to extend the coverage to higher elevations to reduce the cone of silence. It is fair to state that the 30° elevation might be a practical upper limit for the WSR-88D. Top elevations higher than 20° have not been justified by strong meteorological reasons.

2.1.4 Signal to noise ratio

The SNR listed in Table 1 provides the specified accuracy of velocity and spectrum width measurements to the range of 230 km for both rain and snowfall rates of about 0.3 mm of liquid water depth per hour. That is, at a range of 230 km the SNR is larger than 10 dB thus the accuracy of Doppler measurements to shorter ranges is independent of noise and solely a function of number of samples and Doppler spectrum width.

2.1.5 Spatial resolution

The angular resolution is principally determined by the need to resolve meteorological phenomena such as tornados and mesocyclones to ranges of about 230 km, and the practical limitations imposed by antenna size at wavelength of 0.1 m. Even though beamwidth of 1° provides relatively high resolution, the spatial resolution at 230 km is 4 km. Because the beam of the WSR-88D is scanning azimuthally, the effective angular resolution in the azimuthal direction is somewhat larger (Doviak & Zrnic, 2006, Section 7.8); typically, about 40% at the 3 RPM scan rates of the WSR-88D. This exceeds many mesocyclone diameters, and thus these important weather phenomena, precursors of many tornadoes, can be missed. Tornadoes have even smaller diameters and therefore can not be resolved at the 230 km range.

The range resolution is indirectly influenced by the angular resolution; there is marginal gain in having range resolution finer than the angular one. For example better range resolution can provide additional shear segments and therefore improve detection of vortices at larger distance. The range resolution for reflectivity is coarser for two reasons: (1) reflectivity is principally used to measure rainfall rates over watersheds which are much larger than mesocyclones and (2) reflectivity samples at a resolution of 250 m are averaged in range (Doviak & Zrnic, 2006, Section 6.3.2) to achieve the required accuracy of 1 dB.

2.1.6 Precision of measurements

The specified 1 dB precision of reflectivity measurements (Table 1) provides about a 15% relative error of stratiform rain rate (Doviak & Zrnic, 2006, eq 8.22a). This has been accepted by the meteorological community. The specified precisions of velocity and spectrum width estimates are those derived from observations of mesocyclones with research radars. The 8 dB SNR is roughly that level beyond which the precision of velocity and spectrum width estimates do not improve significantly (Doviak & Zrnic, 2006, Sections 6.4, 6.5). But, it is possible that lower precisions can be tolerated and benefits can be derived therefrom. For example, it has been proposed (Wood et al., 2001) that velocity estimates be made with less samples (e.g., by a factor of two) in order to improve the azimuthal resolution. Although this increases the error of the Doppler velocity estimates by the square root of two, the improved angular resolution can increase the range, by about 50% (Brown et al., 2002 and 2005), to which mesocyclones and violent tornadoes can be detected. Therefore in the recently introduced scanning patterns, the data (i.e., spectral moments) are provided at 0.5° increments in azimuth (Section 3.5).

2.2 Radar operation

The essence of the hardware (Fig. 1) is what radar operators see on the console. To the left of the data link (R,V,W,D) is the radar data acquisition (RDA) part consisting of the transmitter, antenna, microwave circuits, receiver, and signal processor. These components are located at radar site and data is transmitted to the local forecast office (LFO) where Radar Product Generation (RPG, Fig. 1) takes place. Operators at the LFO control (the block Control in Fig. 1) the radar and observe/analyze displays of data fields. At a glance of a console they can see the operating status of the radar and data flow. In the RPG the data is transformed into meteorologically meaningful information (Products in Fig. 1) by algorithms executed on Ethernet cluster of workstation.

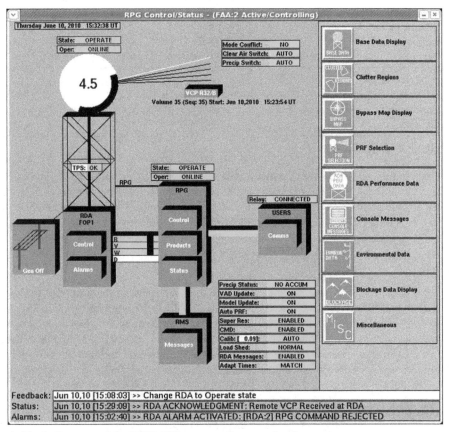

Fig. 1. Block diagram of the WSR-88D seen on the console of operators.

The radar is fully coherent pulsed Doppler and pertinent parameters are listed in Table 2 (see also Doviak & Zrnic, 2006 page 47). Each radar is assigned a fixed frequency in the band (Table 2), hence some values like the beamwidth and unambiguous velocities (not listed) depend on the exact frequency.

Frequency	2.7 to 3 GHz
Beamwidth	1°
Antenna gain	44.5 to 45.5 dB
Transmitter:	
Pulse power	750 kW
Pulse width	1.57 µs and 4.57 µs
Rf duty cycle	0.002
PRFs (Hz, 5 sets of 8, variation ~3%)	322, 446, 644, 857, 1014, 1095, 1181, 1282
Unambiguous range (km)	466, 336, 233, 175, 148, 137, 127, 117
Receiver linear:	
Dynamic range	94 dB at 1.57 µs pulse and 99 dB at 4.57 µs
Intermediate frequency (IF)	57.6 MHz
A/D converter at IF	14 bits
Sampling rate	71.9 MHz
Noise figure	-113 dBm at 1.57 µs and -118 dBm at 4.57 µs
Filter bandwidth or type:	
Front end analogue	6 MHz (3 dB bandwidth)
IF Digital matched, short/long pulse	Output samples spaced at 250 m/500m
Radial spacing in azimuth	1° or 0.5°

Table 2. Radar characteristics.

The data coming out of the RDA consist of housekeeping (time, pointing direction of the antenna, status, operating mode, and fields of reflectivity factor, mean radial velocity, and spread of velocities (designated as R, V, W in the console, Fig. 1), collectively called spectral moments. A wideband communication link is used to exchange base data and radar status/control between RDA and RPG. Depending on distance this link is by direct wire (up to 120 m), microwave line-of-site (to 38 km), or telephone company T1 line (unlimited).

Pulse of high peak power and narrow width (Table 2) generated at the output of the power amplifier is guided to the antenna. It is radiated in form of electromagnetic (EM) field confined within the narrow (1°) antenna beam. The propagating EM field interacts with intervening scatterers (precipitation, biological, and other). Part of the field is reflected forming a continuous stream at the antenna where it is intercepted and transformed for further processing by the receiver. Concise mathematical expression for the magnitude of the electric field at a distance r from the radar is

$$E = \left[\frac{P_a \eta}{\pi}\right]^{1/2} \frac{f(\theta, \phi)}{2r} \cos\left[2\pi f\left(t - \frac{r}{c}\right) + \psi_t\right] U(t - r/c), \tag{1}$$

where P_a is the power radiated by the antenna, r is the distance, $f(\theta,\phi)$ is the antenna pattern function (one way voltage), η is the free space impedance ($120\pi\ \Omega$), c speed of light, f radar frequency, and ψ_t arbitrary phase at the antenna. $U(t\text{-}r/c)$ designates the pulse function such that it is 1 if its argument is between 0 and τ (the pulse width).

2.2.1 Radar signal and Doppler shift

The effective beam cross section and pulse width define the intrinsic radar resolution volume but processing by the receiver increases it in range. Scatterers (hydrometeors such as rain, hail, snow and also insects, birds etc.,) within the resolution volume contribute to the backscattered electric field which upon reception by the antenna is transformed into a microwave signal. The signal is converted to an intermediate frequency f_{if} then passed through anti-alias filter (nominal passband ~ 14 MHz), digitized (as per Table 2), and down converted to audio frequencies (base band) for further processing.

At intermediate frequency the signal coming from a continuum of scatterers can be represented as $A(t)\cos(\omega_{if}t + \omega_d t)$ where the amplitude $A(t)$ fluctuates due to contribution by scatterers and ω_d is the instantaneous Doppler shift caused by their motions toward (positive shift) and/or away (negative shift) from the radar. To determine the mean sense of motion (sign of Doppler shift) the intermediate frequency is removed and the signal is decomposed into its sinusoidal and cosinusoidal components, the inphase I and quadrature phase Q parts. These carry information about the number and sizes of scatterers as well as their motion. Samples of I and Q components are taken at consecutive delays with respect to the transmitted pulse. The delays are proportional to the range within the cloud from which the transmitted pulse is reflected. Samples from the same range locations (delays) are combined to obtain estimates of the spectral moments: reflectivity factor Z, mean Doppler velocity v, and spectrum width σ_v (Doviak & Zrnic, 2006). The Doppler velocity v is related to the frequency shift f_d and wavelength λ via the Doppler equation

$$f_d = 2v/\lambda, \tag{2}$$

and so is the spectrum width.

Radars display (and store) equivalent reflectivity factor (often denoted with Z_e) which is computed from the power and other parameters in the radar equation (Doviak & Zrnic 2006) assuming the scatterers have refractive index of liquid water. For small (compared to wavelength) spherical scatterers, Z_e expressed as function of the distribution of sizes $N(D)$, equals

$$Z_e = \int_0^{D_{max}} N(D)D^6 dD. \tag{3}$$

2.2.2 Processing path from signals to algorithms

Top left part in Fig 2 illustrates the continuum of returns (either I or Q), after each transmitted pulse from 1,...to M. Thus M samples at a fixed range delay (double vertical line) are operated on in various ways to produce estimates. There are as many estimates along range time as there are samples. That is, sample spacing is typically equal to pulse duration and therefore consecutive samples are almost independent. Closer sampling (i.e., oversampling) has some advantages (Section 4.2).

Radials of spectral moments are transmitted to the RPG (a radial of velocities is in the top right part of Fig. 2). Spectral moments are displayed at Weather Forecast Offices, are recorded, and are also processed by algorithms to automatically identify hazardous weather features, estimate amounts of precipitation, and to be used in numerical models among other applications. Example displayed in Fig. 2 (right bottom) is the field of Doppler velocities obtained by the WSR-88D in Dove, North Carolina during the Hurricane Irene on Aug 28th, 2011 at 2:29 UTC. The end range on the display is 230 km which is also the range up to which quantitative measurements are currently being made. Extension to 300 km is planned.

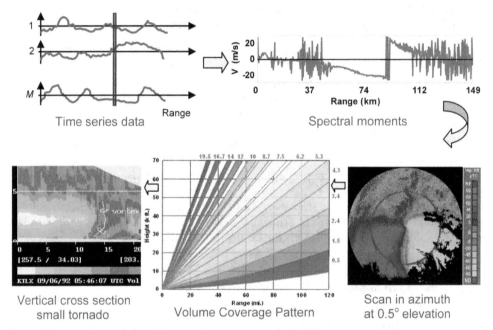

Fig. 2. Information path from time series to output of algorithms.

The radar is sufficiently sensitive to detect precipitation at much larger ranges where the beamwidth and observations high above ground mar quantitative interpretation of impending weather on the ground. At the elevation of 0.5º, the radar makes two scans: one with the longest PRT (3.1 ms) for estimating reflectivities unambiguously up to 465 km in range, the other with one of the short PRTs to estimate unambiguously velocity over a sufficiently large span. The ambiguities in range and velocity are inherent to pulsed Doppler radars. Reflections from scatterers spaced by the unambiguous range ($r_a = cT_s/2$ where T_s is pulse repetition time) appear at the same delay with respect to the reference time (determined by the last of two transmitted pulse). Obvious increase in range can be made by increasing T_s. And this is fine for measurements of reflectivity but would harm measurements of velocity. At the 10 cm wavelength Doppler velocities are estimated from the change in phase of the returned signal (Doviak & Zrnic 2006). Thus the WSR-88D is a phase sampling and measuring instrument. The change in phase of the return from one pulse to the next $2\pi f_d T_s$ is proportional to the Doppler velocity v as indicated in (2).

If the phase change caused by precipitation is outside the $-\pi$ to π interval it cannot be easily distinguished from the change within this interval. These limits define the unambiguous frequency $f_a = \pm 1/(2T_s)$ and through the Doppler relation (2) the unambiguous velocity as

$$v_a = \lambda/(4T_s). \tag{4}$$

Scatterers do cause a Doppler shift within the pulse as it is propagating and reflecting, but this shift is very small and can not be measured reliably as the following argument demonstrates. Consider the $\tau = 1.57$ μs pulse width (WSR-88D) and scatterers moving at 10 m s^{-1} (36 km h^{-1}). The corresponding Doppler frequency shift is 200 Hz (at 10 cm wavelength) and it produces a phase difference of 0.11° $(2\pi f_d \tau)$ between the beginning and end of the pulse return. This tiny difference can not be measured with sufficient accuracy to yield useful estimate.

To mitigate the ambiguity problem the WSR-88D has some options one of which is special phase coding and processing. The result is seen in Fig. 2 where the pink ring at 137 km indicates the unambiguous range for velocity measurements (see discussion in section 3.2.3); it represents censored data because the ground clutter from nearby range and weather signals from the second trip range are comparable in power and can not be reliably separated.

Operators of the WSR-88D have at their disposal preprogrammed volume coverage patterns (VCP – see example in Fig. 2). These are consecutive scans starting from elevation of 0.5° and incrementing until a top elevation is reached. Most algorithms require a full volume scan to generate a product. The one in Fig. 2 (bottom left) reconstructs a vertical profile of Doppler velocities along a radial; the radar is located to the right and green colors indicate velocities toward the radar in 5 m s^{-1} increments starting with 0 (gray color). Cylindrical protrusion below 5 km in the middle with some velocities toward the radar (red color) is indicative of a tornado.

3. Signal processing and display

The block diagram (Fig. 3) of the WSR-88D radar is typical for pulsed Doppler radars. Essential components are the Frequency and Timing generator, the transmitter and the receiver. Radar and antenna controls are omitted from the figure. Intermediate frequency (if) on the radars is 57.6 MHz, and the local oscillator (lo) frequency is adjustable to cover the range between 2.7 and 3 GHz (the operating band, see Table 2). The power amplifier is a klystron. The transmit/receive switch is comprised of a circulator and additional devices to protect the receiver from the transmitted high power pulse. The low noise amplifier (LNA) has a noise figure ~ 0.8 dB and the receiver bandwidth is 6 MHz up to the input of the digital receiver. The digital receiver is a proprietary product of SIGMET Co (now Vaisala) and its essence is described next.

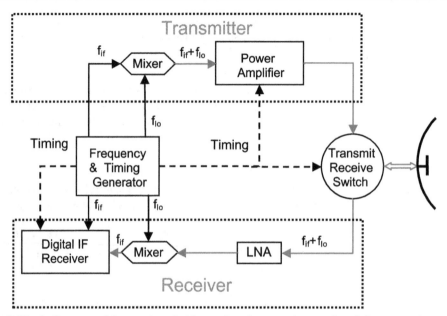

Fig. 3. Block diagram of the receiver (without signal processing part) and the transmitter.

3.1 Digital receiver

The analogue signal $A(t)\cos(\omega_{if}t + \omega_d t)$ is sampled at a rate of 71.9 MHz producing a stream (time t_i) of 14 bit numbers. These are multiplied (Fig. 4) with $\sin(\omega_{if}t_i)$ and $\cos(\omega_{if}t_i)$ and digitally filtered to obtain the base band I and Q components (at times t_k). Although the nominal short pulse duration is 1.57 μs same as sample spacing in range, 155 samples spaced at ~ 13.8 ns over 2.15 μs interval are used for multiplication and filtering (in the long pulse mode the number of samples is 470 over a 6.53 μs interval). The digital low pass filter is adjusted to match the shape of the transmitted long or short pulse. Matching is achieved by passing the attenuated transmitted pulse ("burst") through the receiver and taking the discrete Fourier transform of the output. The inverse of this transform gives the coefficients of the matched impulse response filter. Amplitude and phase of the "burst" is sampled upon each transmission to monitor power, compensate for phase instabilities, and use in phase codes for mitigating range ambiguities. The timing diagram (Fig. 5) illustrates the relations between transmitted sequence, digital oscillator samples, the sampled sequence from a point scatterer and its I and Q values (after the matched filter).

3.2 Transmitted sequences and volume scans

Several volume coverage patterns are available. With the exception of one all utilize the short pulse. The exception has a uniform sequence of long pulses at the longest PRT for observations in clear air or snow where weak reflections are from insects, birds, ice and/or refractive index fluctuations. For storm observations the volume coverage patterns have three distinct modes depending on the elevation.

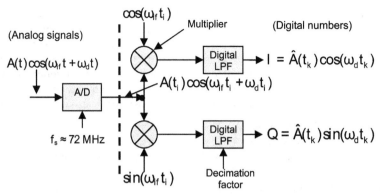

Fig. 4. Conceptual diagram of the digital receiver and down converter indicating the essential operations. The dashed vertical line shows where digital processing begins.

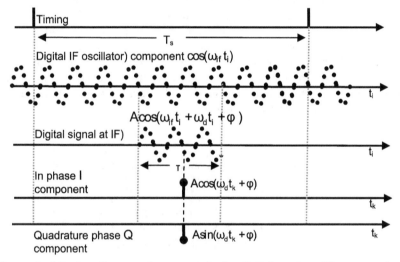

Fig. 5. Conceptual timing diagram of processes in the digital receiver. The return signal is assumed to be sinusoidal pulse such as would be produced by a single point scatterer.

3.2.1 Lowest elevation scans

At the lowest two (sometime three) elevations (< 1.6°) two consecutive scans at each elevation are made. For surveillance and reflectivity measurement the longest PRT is used so that the unambiguous range is ~460 km. It is followed by one or more of the higher PRTs for measurement of Doppler velocity and spectrum width whereby the unambiguous velocity interval is larger than ~ 20 m s^{-1}. Thus Doppler estimates can be ambiguous and overlaid in range. To determine the location of the Doppler estimates, powers along the radial at the same azimuth but in the surveillance scans are examined. The echoes from ranges spaced by nPRTc/2 of the Doppler scan, where n is 1, 2, 3, 4, can be overlaid in the Doppler scan; the echo for n=1 is said to come from the first trip because it corresponds to the round trip shorter than the separation between consecutive pulses. Powers from locations spaced by $n\text{PRT}c/2$ are compared to determine the correct range of the Doppler estimates and presence of overlaid echoes. If one of the overlaid powers is larger than user specified threshold (typically 5 dB) the corresponding Doppler spectral moment is assigned to the correct range whereas the values at location of the other overlaid echoes are censored. If the powers are within 5 dB, the variables at all locations where the overlay is possible are censored. Because the Doppler spectral moments are computed and recorded only to the distance of at most twice the unambiguous range the censoring is also done to that distance.

There is a special VCP (Zittel et al., 2008) with three scans at same elevation on five consecutive lowest elevations whereby velocities from three PRFs (No. 4, 6, and 8 in table 2) are combined to increase the v_a and display it up to the distance of 175 km.

3.2.2 Scans at mid and high elevations

At elevations between 1.6° and 7° a "batch" sequence is transmitted. It is a dual PRF in which the first few (3 to 12) pulses are at the lowest PRF and the rest (between about 25 and 90) are transmitted at one of the four highest PRFs (shortest PRTs, Table 2). The lowest PRF pulses are for surveillance, reflectivity measurements, and censoring and assignment of range to Doppler spectral moments; just the same as in the lowest scans. To improve accuracy of the reflectivity estimates powers from the Doppler sequence (high PRF) are included in the averaging provided there is no contamination by overlaid echoes. Beyond 7° elevation uniform PRTs are transmitted because the tops of storms at locations where overlay can occur are below the radar beam.

3.2.3 Phase coding

To mitigate range overlay some volume scanning patterns at the lowest elevations (<2°) have transmitted sequences encoded with the SZ(8/64) phase code (Sachidananda & Zrnic, 1999). The concept is depicted in Fig. 6 and explained in the caption. The prescribed phases Ψ_k (i.e., switching phases) are applied to the transmitted pulses. Formally this is represented by multiplication of the sequence with the switching code $a_k = \exp(j\Psi_k)$. The first trip return signal is made coherent by multiplying it with the conjugate $a_k^* = \exp(-j\Psi_k)$. With this multiplication the 2nd trip signal is phase modulated by the code $c_k = a_{k-1}a_k^*$. The 2nd trip can be made coherent by multiplying the incoming sequence with a_{k-1}^*, in which case the 1st trip signal is modulated by the code c_k^*. The code, a_k is designed such that the modulation code c_k has a phase shift given by $\varphi_k = \Psi_{k-1} - \Psi_k = 8\pi k^2 / 64$. The special property of this code is that its autocorrelation is unity for lags in multiples of 8 (lags $8n$; $n=0,1,2,...$), and is zero for all other lags. Therefore the power spectrum has only 8 non-zero coefficients separated by $M/8$ coefficients. The SZ(8/64) switching code is given by

$$a_k = \exp[-j\sum_{m=0}^{k}(\pi m^2 / 8)]; \quad k = 0,1,2...63. \tag{5}$$

It has periodicity of 32 hence the number of samples M must be an integer multiple of 32. From (5) it is obvious that the phase sequence consists of a binary sub multiple of 360° hence it is generated without round-off errors using standard binary phase shifters. Because the desired phase and actual phase might not be exactly equal, the transmitted phase is sampled and used in processing to precisely cohere the signal from the desired trip.

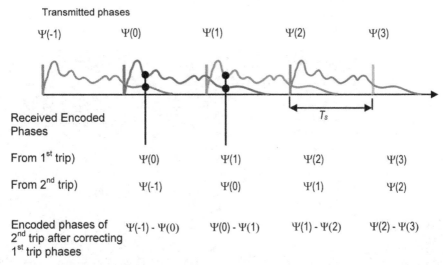

Fig. 6. Transmitted pulse sequence (vertical color lines) and the corresponding received powers (wiggly curves). The phases of the transmitted pulses are indicated and indexed from -1 to 3. The location of overlay at one fixed range is indicated by two black vertical lines. The phases of the received signals from the first and second trip are indicated as well as the phase of the second trip signal after subtracting (correcting) the phase of the first trip.

In case of overlaid echoes the phase coding allows separation of the contributions by the first and second trip signals. This is accomplished by first cohering (correcting) the phases of the stronger echo, then filtering it out. For example if first trip is cohered the second trip signal spectrum (complex with magnitudes and phases) is split into eight replicas over the unambiguous interval. Then frequency domain filtering of the first (strong) trip signal with a notch centered on its spectrum and having a width of ¾ unambiguous interval leaves two spectral replicas of the second trip signal spectrum. From these replicas it is possible to reconstruct the second trip spectrum and compute spectral moments. It turns out that cohering for the first trip signal induces 4 spectral replicas in the third trip signal and again eight replicas into the fourth trip signal; the fifth trip signal has two replicas and can not be recovered.

Determination of the ranges where overlaid echoes might be is made using powers from the surveillance scan (long PRT) which precedes the Doppler scan (phase coded short PRT). The overlay trip number and powers are needed to make proper cohering-recohering order and notch filter application. In case ground clutter is present Blackman window is applied to time series data and clutter is taken out with a special frequency domain filter (Sec 3.3). If there is no clutter contamination but overlaid echoes are present the von Hann window is chosen. An example of Doppler velocity fields obtained with the SZ(8/64) phase code is in Fig. 7 (left side). The same field obtained by processing and censoring with no phase coding is also plotted (right side); note the large pink area in the second trip region indicative of non recoverable velocities. Small pink areas in the first trip region (SE of radar) signify that overlaid powers of first and second trip signals are within 10 dB and hence velocities can not be confidently recovered. There is a narrow pink ring of censored data in the image where phase code is applied. The beginning range of the ring is at the start of the second trip (175 km) and is caused by automatic receiver shut down during transmission followed by the strong first trip ground clutter overwhelming the weaker second trip signal.

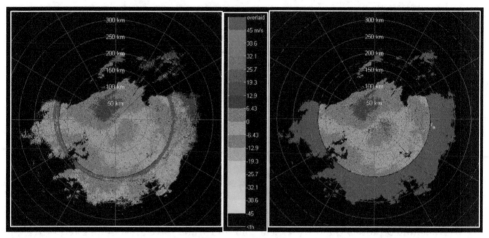

Fig. 7. Fields of Doppler velocities. Left: obtained from phased coded sequence. Right: obtained from non coded sequences. Elevation is 0.5°, the unambiguous velocity v_a= 23.7 m s^{-1} and range r_a = 175 km. Data were collected on 10/08/2002 with the research WSR-88D (KOUN). The color bar indicates velocities in m s^{-1}. (Figure from Torres et al., 2004c).

3.3 Ground clutter filter

The ground clutter filter implemented on the network is a frequency domain filter with interpolation over the removed clutter spectral coefficients. The filter called Gaussian Model Adaptive Processing (GMAP) has been developed by Siggia and Passarelli (2004). Its first premise is: clutter has a Gaussian shape power spectrum with width linearly related to the antenna rotation rate; hence the width can be computed. The second is the signal spectrum has also Gaussian shape and has width larger than clutter's. The Blackman window is applied followed by Fourier transform. Receiver noise is externally provided to the filter and used to establish the spectral noise level which helps determine how many spectral coefficients either side of zero to remove (Fig. 8, blue peak is from ground clutter). The removed coefficients are replaced (iteratively) with a Gaussian curve obtained from Doppler moments and the spectrum of the weather signal (dotted curve) is restored. Then the inverse discrete Fourier transform is performed to obtain the autocorrelation at lag 1. The argument of the autocorrelation is linearly related to the mean Doppler velocity (see section 3.4).

Several options exist to decide where to filter clutter. One relies on the clutter map to locate azimuths and ranges. It is also possible but undesirable to apply clutter filter everywhere. The operators can select regions between azimuths and ranges where to turn the filter on. Recently an adaptive algorithm called Clutter Mitigation Decision has been implemented (Hubbert et al., 2009). It uses coherency of the clutter signal exemplified in what the authors call Clutter Phase Alignment (CPA) defined as $CPA = |\operatorname{sum} V_k| / \operatorname{sum} |V_k|$, where V_k is the complex voltage (I + j Q) from a fixed clutter location at consecutive times (spaced by the PRT) indicated by time index k and the sum is over the total number of pulses in the dwell time. Local standard deviation (termed texture) of reflectivity factor Z_i in range (i index indicates adjacent values in range) and changes in sign of the differences $Z_{i+1} - Z_i$ are also used; the frequency of change in reflectivity gradient along range is obtained from this difference and it defines the spin variable. The CPA, texture, and spin are combined in a fuzzy classification scheme to identify locations where clutter filter should be applied.

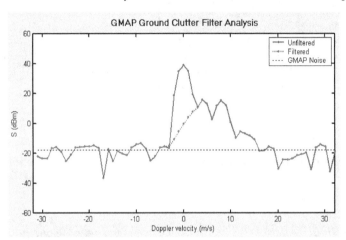

Fig. 8. Doppler spectrum of simulated weather signal (red) and clutter (blue). Interpolated (filtered) Gaussian part and estimated noise level are shown. The v_a=32 m s^{-1}. (Figure as in Torres et al., 2004c).

The GMAP filter and censoring (Free & Patel, 2007) is applied to surveillance and Doppler scans. In the "batch" mode the number of samples is insufficient for spectral processing hence the average voltage (i.e., DC) from the samples spaced by the long PRT is removed.

The system also employs strong point clutter (typically caused by aircraft) removal along radials. It is done on each spectral moment independently by comparing the sample power with two adjacent values either side of it. If the value is outside prescribed criteria it is replaced by interpolation of neighboring values.

3.4 Computation of spectral moments

In computations of Z and σ_v receiver noise powers are subtracted from the returned powers. Thus, the receiver noise power is estimated at the end of each volume scan at high elevation angle. The noise depends on the elevation angle because contributions from ground radiation and air constituents are larger if the beam is closer to the ground. To account for the increase the noise is extrapolated to lower elevations using empirical relations.

The reflectivity factor is obtained by summing the pulse powers, subtracting the noise power, and using the radar equation (Doviak & Zrnic, 2006). At the lowest few elevations Z is computed from the long PRT (surveillance scan). At mid elevations ("batch mode") the reflectivity is computed from both the long and short PRTs if no overlay is indicated; otherwise only samples from the surveillance scan (long PRT) are used.

Computation of Doppler variables starts with the discrete Fourier transform. In absence of clutter, time series data is equally weighted (uniform window) and the power spectrum estimate (at some range location) is

$$\hat{S}(k) = \left| \frac{1}{M} \sum_{m=0}^{M-1} V(m)e^{-j\frac{2\pi mk}{M}} \right|^2, \quad k = 0,1,...,M-1 \tag{6}$$

The discrete inverse Fourier transform applied to (6) produces the value of circular autocorrelation function at lag 1 (i.e., T_s) which contains one erroneous term, namely the product of first and last member of the time series (Torres et al., 2007). This term is subtracted so that the autocorrelation at lag one (i.e., T_s) becomes

$$\hat{R}(1) = \sum_{m=0}^{M-1} \hat{S}(k)e^{j\frac{2\pi k}{M}} - \frac{1}{M}V^*(M-1)V(0), \tag{7}$$

and the mean velocity estimate comes out to be (Doviak & Zrnic, 2006. eq 6.19)

$$\hat{v} = -(\frac{\lambda}{4\pi T_s})\arg[\hat{R}(1)]. \tag{8}$$

The spectrum width for most VCPs is estimated by combining the lag one autocorrelation and the signal power $\hat{P}_s = \sum_{m=0}^{M-1} |V(m)|^2 - P_n$, from which the noise power P_n is subtracted, as follows (Doviak & Zrnic 2006, eq 6.27)

$$\hat{\sigma}_v = \frac{\lambda}{2\sqrt{2}\pi T_s} \left| \ln\left(\frac{\hat{P}_s}{|\hat{R}(1)|} \right) \right|^{1/2}. \tag{9}$$

But, if the logarithm term is negative $\hat{\sigma}_v$ is set to zero. In case of phase coding and presence of overlaid echoes equation (9) is used for the weaker signal in the surveillance scan (long PRT). The spectrum width of the strong signal is computed for the Doppler scan using the ratio $\hat{R}(1)/\hat{R}(2)$ as in Doviak & Zrnic (2006, eq. 6.32), because it is not biased by presence of the weak signal.

3.5 Oversampling in azimuth (overlapping radials)

Until recent upgrades all VCPs had spacing of radials at 1º azimuth and reflectivities were averaged and recorded at 1 km range intervals but velocities retained inherent spacing of 250 m (Table 1). Newly added VCPs employ a strategy whereby at the lowest two elevations time series data from overlapping (in azimuth) beams are processed to produce spectral moments. Thus data obtained over one degree azimuth are weighted with the von Hann window and so are data from the adjacent azimuth centered 0.5° off from the previous. This produces more radials of data (spaced by 0.5° as opposed to 1°) increasing resolution to facilitate recognition of small phenomena such as tornado vortices (Brown et al., 2002, and 2005). The contrast between the routine and enhanced resolution of a tornado vortex signature is evident in the example in Fig. 9. The reflectivity field (top figures in dBZ as indicated by the color bars) displays a "hook echo" associated with low level circulation. The crisp pattern (top right) is the result of the enhanced resolution.

The velocity field (bottom in Fig. 9) displays three circular features ("balls") in its center: the lighter green and red adjacent to it in azimuth indicate cyclonic circulation (mesocyclone). Its diameter is about four km and it is estimated from the distance between maximum inbound (green) and outbound (red) velocities. The sharp discontinuity in the center (light green ~ -30 m s^{-1} to > 30 m s^{-1}) is the tornado vortex signature (TVS). The transition between the red "ball" and the green one farther in range marks the zero radial velocity suggesting converging flow (i.e., red and green velocities pushing air toward each other) near ground. Bottom right: same as in the left but the resolution in azimuth is enhanced to 0.5°. The TVS is better defined and so are other small scale features.

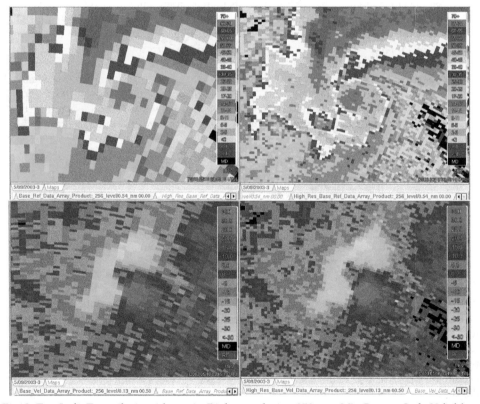

Fig. 9. Top Left: Z, resolution 1 km x 1°. Right: resolution 250 m x 0.5°. Bottom Left: V field, resolution 250 m x 1°. Right: resolution 250 m x 0.5°. X, Y sizes are 25 by 20 km; radar is at x= 4 km and y = -25 km with respect to each image left corner. (Courtesy, S. Torres).

4. Near term enhancements

Currently a significant transformation of the radars is ongoing; it is addition of dual polarization (Zrnic et al., 2008). By mid 2013 all radars on the network should have this capability. Although Doppler capability is not a prerequisite for dual polarization, the coherency of transmit-receive signals within one PRT is for differential phase measurement. Dual polarization offers ample possibilities for application of spectral analysis to polarimetric signals and these are being explored (e.g., to discriminate between insects and birds, Bachman & Zrnic, 2007; to suppress ground clutter, Unal, 2009; or to achieve adaptive clutter and noise suppression, Moisseev & Chandrasekar, 2009).

Three improvements approved for soon inclusion on the network are pending. These are staggered PRT, processing of range oversampled signals, and adaptive recognition and filtering of ground clutter. Brief description follows.

4.1 Staggered PRT

It is planned for mitigating range velocity ambiguities at mid elevation angles with possible use at the lower elevations. The scheme consists of alternating interval between transmitted pulses (Fig. 10) and estimating arguments of two autocorrelations at the two lags, $\arg[R(T_1)]$ and $\arg[R(T_2)]$. The velocities estimated from these arguments have a different unambiguous interval (each inversely proportional to the corresponding separation T_i, i=1 or 2) as can be deduced from eq. (8). Therefore the difference of the velocities uniquely tags the proper unambiguous interval for either PRT so that correct dealiasing can be achieved (Torres et al., 2004a) up to larger v_a than possible with only one of these PRTs . For the example in Fig. 10, $v_a = 3v_{a2} = 2v_{a1}$. Consider T_1=1 ms T_2=1.5 ms which produces v_a = 50 m s^{-1} (unambiguous interval is -50 to 50 m s^{-1}) and unambiguous range of at least 150 km.

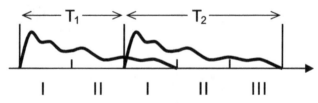

Fig. 10. Staggered PRT. The stagger ratio T_1/T_2 = 2/3. The continuous curve depicts the return from precipitation extending up to $cT_2/2$ but not further (from Torres et al., 2009 and adapted from Sachidananda & Zrnic, 2003).

Power estimates in range sections I, II, and III (Fig.10) are computed separately for the short PRT and the long PRT to check if data censoring is needed. Comparison of powers in the two PRT intervals indicates if there is overlay and how severe it is so that appropriate censoring can be applied. In Fig. 11 contrasted are two fields of velocities obtained with two radars (spaced about 20 km apart). The left field comes from the operational WSR-88D in Oklahoma City and was obtained with the "batch mode" and parameters as indicated. On the right is the same storm complex but obtained with staggered PRT on the research WSR-88D radar in Norman OK some 20 km SSW from Oklahoma City. Highlighted in yellow circles are regions where significant aliasing occurs on the operational radar (exemplified by abrupt discontinuities in the field, change from red to green) but are absent in the field from the research radar. Also, the large pink area of overlaid echoes has almost disappeared in the measurement made utilizing the staggered PRT. The small circle closest to the radar origin indicates overlaid echo contaminating the first trip velocities of the operational radar.

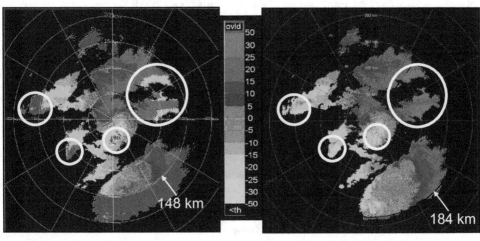

$$V_{a1} = 25.4 \text{ m s}^{-1} \qquad\qquad V_a = 45.2 \text{ m s}^{-1}$$

Fig. 11. Velocity fields of a storm system. Left: field obtained with the operational WSR-88D radar in Oklahoma City on April 06, 2003, elevation 2.5 deg, batch mode with unambiguous range of 148 km and velocity of 25.4 m s^{-1}. Pink regions locate censored velocities which can not be reliably recovered due to overlaid first and second trip echoes. Right: same as on the left but obtained with the research WSR-88D (KOUN) utilizing staggered PRT. This radar is about 20 km south from the operational radar. The color bar indicates velocities (m s^{-1}), red away from and green toward the radar. (Figure adapted from Torres et al., 2003).

4.2 Oversampling techniques

Oversampling here indicates spacing of I, Q samples smaller than the pulse duration. Operations on few of these range consecutive "oversamples" can reduce error in estimates and/or data acquisition time (Torres & Zrnic, 2003). Simplest of operations is averaging in range of oversampled spectral moments. Somewhat more involved is the whitening transformation in which the signal vector $\mathbf{v} = [V(m,0), V(m,1),... V(m, l),....V(m, L)]$ consisting of L oversampled correlated complex voltages is transformed into a set of L orthogonal voltages (Torres & Zrnic, 2003). The time index m refers to the usual sample time and l to the oversampled range time. The transformation takes the form $\mathbf{x} = \mathbf{H}^{-1}\mathbf{v}$ with $\mathbf{H}$ related to the normalized correlation matrix $\mathbf{C}$ of $\mathbf{v}$ via $\mathbf{C} = \mathbf{H}^*\mathbf{H}^T$. The correlation matrix can be pre-computed (or measured e.g., Ivic et al., 2003) because it depends solely on the envelope of the transmitted pulse and the baseband equivalent receiver filter shape for a uniform Z. The L transformed samples are independent and averaging of spectral moments obtained from each (in absence of noise) yields smaller error of estimates. Whitening is effective at large SNRs but fails otherwise. To achieve L independent samples the receiver filter bandwidth needs to be increased L times over the matched filter bandwidth and this enhances the noise by the same factor. In addition the whitening transformation also increases the noise hence the net SNR reduction is proportional to L^2. Practical L is about 3 to 6, so the decrease is not catastrophic considering that weather SNRs are mostly larger than 20 dB. Another issue concerning whitening is the shape of the range weighting function compared to the matched filter. The two weighting functions have the same range extent but the one from whitening has rectangular shape smearing slightly its range resolution.

Increasing the number of independent samples when it is advantageous and gradually reverting to the matched filter has also been proposed (Torres et al. 2004b) and implemented (Curtis & Torres, 2011) on the National Weather Radar Testbed (NWRT), a phased array radar antenna powered by a WSR-88D transmitter (Zrnic et al., 2007). The processing is called adaptive pseudowhitening. It requires initial estimates of SNR and spectrum width.

Vivid example contrasting adaptive pseudowhitening to standard processing illustrates the much smoother fields obtained with the former (see Fig.12, and caption). The gradient of Doppler velocities (indicated with an arrow) is at the interface of the storms outflow and the environmental flow. This type of discontinuity is the key feature detected by algorithms for locating gust fronts and quantifying wind shear across the boundary; such information is extremely useful for air traffic management and safety at airports.

In contrast to whitening techniques pulse compression does not degrade the SNR (Doviak and Zrnic, 2006) but is not considered due to excessive bandwidth and current hardware constraints. A very simple alternative to speed volume coverage at lowest elevations (where tornadoes are observed) is a VCP with adaptive top elevation angle based on radar measurements (Chrisman et al., 2009). It will soon be added to the VCPs on the network.

4.3 Clutter detection and filtering

A novel way to recognize and filter ground clutter is planned. Its acronym CLEAN-AP stands for clutter environment analysis using adaptive processing (Warde & Torres, 2009). The essence of the technique is spectral analysis (decomposition) of the autocorrelation at lag 1 and use of its phase at and near zero Doppler shift. The conventional estimate

$$\hat{R}_b(1) = \frac{1}{M^2} \sum_{0}^{M-1} |Z(k)|^2 e^{j2\pi k/M} , \qquad (10)$$

where $Z(k)$ is the discrete Fourier transform of the returned signal, is biased (indicated by subscript b) and can be unbiased as in (7). Another way to avoid the bias is by computing two Fourier transform as proposed by (Warde & Torres 2009). One, $Z_0(k)$ is the complex spectrum of $d(m)V(m)$, $d(m)$=window function, and the other $Z_1(k)$ is the spectrum of $d(m)V(m+1)$ from the sequence shifted in time by one unit (T_s). Then the unbiased estimate is

$$\hat{R}(1) = \sum_{k=0}^{M-1} Z_0^*(k) Z_1(k) / M^2. \qquad (11)$$

Individual terms $S_1(k) = Z_0^*(k) Z_1(k)$ constitute the spectral density (over Doppler index k) of the lag 1 autocorrelation function. Thus the autocorrelation spectral density is estimated in CLEAN-AP from the cross spectrum.

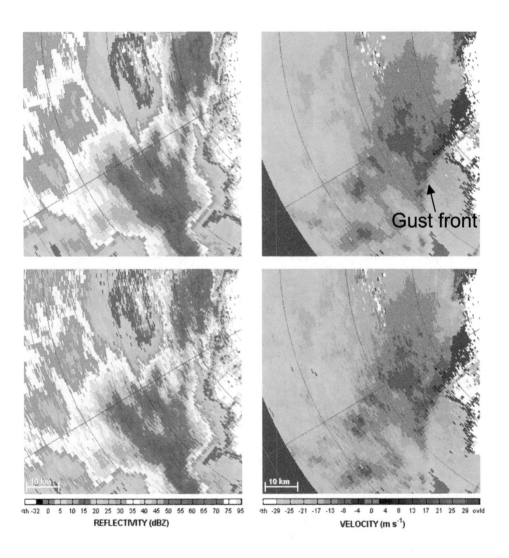

Fig. 12. Fields of reflectivity and velocity from a severe storm obtained on 2 Apr 2010 10:54 UTC, with the phased array radar (NWRT) in Norman, OK. Top two panels resulted for pseudowhitening applied to $L = 4$ samples of time series data; the number of samples M per radial was 12 for Z and 26 for v. Data in the lower panels have been obtained by processing as on the WSR-88D (16 for Z and 64 for v). The curved discontinuity in the velocity field delineates outflow boundary (gust front) generated by this storm. The peak reflectivity values of $\sim$ 65 dBZ are likely caused by hail. (Adapted from Curtis & Torres, 2011).

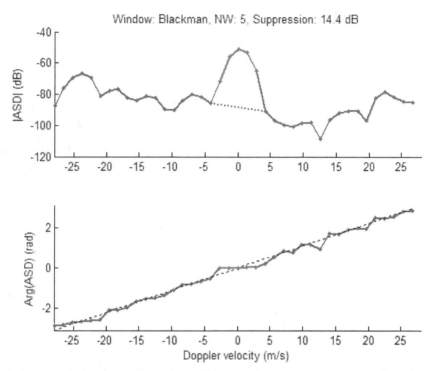

Fig. 13. Autocorrelation spectral density (ASD) of a radar return, top: magnitude and bottom: phase. Clutter is well defined with its peak at zero and flat phase (red). Based on this phase five coefficients are replaced with interpolated values resulting in 14.4 dB of suppression (defined as the ratio of total S+C power to remaining power). Interpolated powers are indicated by the dotted line; dash line represents linear phase; v_a= 27 m s^{-1}. Data obtained with the phased array radar (NWRT). (Figure courtesy of Sebastian Torres).

The choice of window function $d(m)$ is very important because its sidelobes limit the amount of power that can be filtered. The clutter power is computed from the sum of $V(m)$ to obtain the clutter to noise ratio (CNR). Then the CNR is compared with the peak to first sidelobe level (PS$_w$) ratio of four windows (w=rectangular, von Hann, Blackman, and Blackman-Nuttall) and the window whose PS$_w$ exceeds the CNR by the smallest amount is chosen. That way the leakage of the clutter signal away from zero will be below the noise level, while the notch width will be smaller than the one for the other windows satisfying the condition PS$_w$>CNR.

Data windows spread the phase of clutter's $S_1(k)$ either side of zero (k=0) Doppler (Fig. 13). Recognition of the flat phase identifies clutter's presence. Doppler index at which the phase begins to depart from zero (according to a set of criteria) defines the clutter filter width. In the mean the autocorrelation spectral density of noise has linear phase as seen in Fig.13 but semi coherent signals have flattened phases in the vicinity of their mean Doppler shifts. Panels in Fig. 14 demonstrate qualitatively performance of this clutter mitigation technique and the caption highlights results.

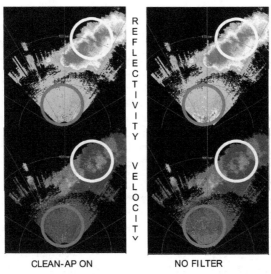

CLEAN-AP ON   NO FILTER

Fig. 14. Fields of reflectivities (top) and velocities (bottom) with no filter (right) and after application of the CLEAN-AP. Close to the radar the strong reflectivities in the top right panel encircled in red (red indicates > 50 dBZ) are caused by ground clutter which also biases the velocities toward zero (lower right panel). CLEAN-AP eliminates most of the clutter in both fields (left panels). To the NE within the yellow circle there are areas of near zero velocities (lower panels gray areas are velocities within ±5 m s⁻¹). These appear unaffected by the filter. The data were collected with the agile beam phased array radar (NWRT) in Norman, OK. (Figure adapted from Warde & Torres, 2009).

4.4 Hybrid spectrum width estimator

The spectrum width estimator (9) is deficient at narrow widths where significant bias occurs. This shortcoming will be overcome with the Hybrid estimator which chooses an appropriate equation depending on a rough initial estimate of σ_v (Meymaris et al., 2009). Initial estimate of the spectrum width is made using thee estimators) (9), $\hat{R}(1) / \hat{R}(2)$ as in (Doviak & Zrnic, 2006 eq. 6.32) and an estimator based on $\hat{R}(1)$, $\hat{R}(2)$, and $\hat{R}(3)$. Criteria applied to the results produce three categories of widths, large, medium, and small. Then (9) is used as estimate for the large category, $\hat{R}(1) / \hat{R}(2)$ for the medium and $\hat{R}(1) / \hat{R}(3)$ for the small.

5. Observations of phenomena

Mesocylone refers to a rotational part of storm with the diameter of maximum wind typically between 3 and 10 km. It is depicted with a couplet of Doppler velocity features (see Fig. 9). Storms having mesocyclones can produce devastating tornadoes (Fig. 9 exhibits a tornado vortex signature associated with the mesocyclone), strong winds, and hail. Thus, much effort has been devoted to detecting and quantifying these phenomena (No. 2 issue of Weather and Forecasting, 1998). One of the motivating reasons for installing Doppler radars in the USA was the potential to detect mesocyclones and tornadoes. The investment in this technology paid off as demonstrated by the graph in Fig. 15. Trend of improvement is seen on all three performance indicators with the steepest rise in the years the Doppler radar network (NEXRAD) was being installed. This is logical: as the new tool was spreading across the country more forecasters were beginning to use it. Improvement continues few years past the completion of the network likely because it took time to train all forecasters and gain experience with the Doppler radar. The data indicates a plateau from about 2002 until present suggesting maturity of the technology with little room left for significant advancements. Further progress might come from combining radar data with short term numerical weather prediction models and/or introduction of rapidly scanning agile beam phase array radars (Zrnic et al., 2007 and Weber et al., 2007).

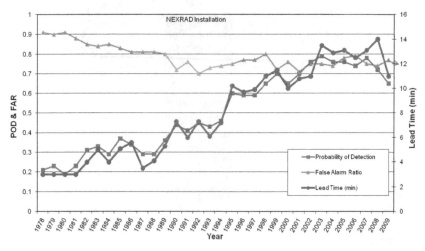

Fig. 15. Probability of detection, false alarms and lead time in tornado warnings issued by the National Weather Service as function of year. (Figure courtesy Don Burgess).

Doppler velocities are potent indicators of diverging (converging) flows such as observed in strong outflows from collapsing storms. These "microbursts" have been implicated in several aircraft accidents motivating deployment of terminal Doppler weather radars (TDWR) at forty seven airports in the USA (Mahapatra, 1999, sec 7.4). Vertical profiles of reflectivity and Doppler velocity in Fig. 16 indicate a pulsing microburst; the intense reflectivity core (red below 5 kft) near ground is the first precipitation shaft and the elongated portion above is the following shaft. On the velocity display the yellow arrows indicate direction of motion. Clear divergence near ground and at the top of the storm (in the anvil) is visible and so is the convergence over the deep mid storm layer (5 to 14 kft). The horizontal change in wind speed near ground of ~ 20 kts at this stage is not strong to pose treat to aviation (35 kts is considered significant for light aircraft).

An atmospheric undular bore (Fig. 17) was observed with the WSR-88D near Oklahoma City. This phenomena is a propagating step disturbance in air properties (temperature, pressure, velocity) followed by oscillation. Spaced by about 10 km the waves propagate in a surface-based stable layer. The layer came from storm outflow and the bore might have been generated by subsequent storm. From the vertical cross section of the velocities it is evident that the positive velocity perturbation (toward the radar) ends at about 4000 ft, above which the ambient flow (green color) resumes. The velocities measured by the radar can quantify the structure of the perturbation, tell the thickness and wavelength. Propagation speed can be estimated by tracking the wave position in space and time.

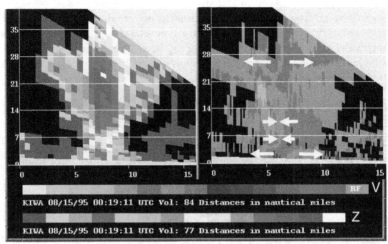

Fig. 16. Vertical cross sections of reflectivity field (left) and Doppler velocity field through a microburst reconstructed from conical scans (up to 19.5° elevation) of the WSR-88D radar in Phoenix Az on Aug 15, 1995. Height is in kft and distances are in nautical miles. The radar is located to the right of each cross section (at about 26 nautical miles). The top color bar depicts velocity categories in non linear increments with red away from the radar: light red = 0-5 kts, dark red 5 to 10, next 10-20; green indicates toward the radar in categories symmetric to red. The bottom bar refers to reflectivities starting at 0 dBZ in steps of 5 dBZ (white category indicates values larger than 65 dBZ).

Fig. 17. Doppler velocities at 0.5° elevation and superposed vertical cross sections of the velocities obtained with Oklahoma City radar on Aug 10, 2011. Red color indicates motion away and green toward the radar located ESE of the bottom right corner. Height lines are in kft above ground level.

Doppler radar is valued for measuring winds in hurricanes and detecting tornadoes that can be imbedded in the bands. Combined with polarimetric capability, its utility greatly increases because of improved quantitative measurement of rainfall. Observation of hurricane Irene which swept the US East coast at the end of August 2011 is the case in point. Rotation speed of over 110 km h^{-1} is apparent in Fig. 18 where the color categories are too coarse to estimate the maximum values. The cyan color captures well Irene's rotational winds because they are aligned with radials. Color categories are coarse precluding precise estimation of velocities but recorded values are quantized to 0.5 m s^{-1}. Although the unambiguous velocity is ~ 28 m s^{-1} values more negative than -30 m s^{-1} are displayed. These and other outside the unambiguous interval have been correctly dealiased by imposing spatial continuity to the field.

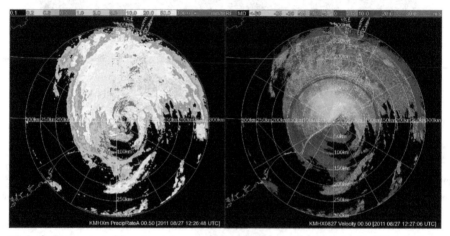

Fig. 18. Left: Rain rate in Hurricane Irene, obtained with a polarimetric algorithm using differential phase and reflectivity factor (surveillance scan with unambiguous range of ~ 465 km). Right: Velocity field obtained with the SZ(2) phase code (Doppler scan with unambiguous range of ~135 km and velocity ~ 28 m s^{-1}). Elevation is 0.5°, time 12:26 UTC, on Aug 27, 2011. The range circles are spaced 50 km apart. Color categories for rain rate are in mm h^{-1} and for velocity in m s^{-1}. (Figure courtesy of Pengfei Zhang).

The rain rate field depicts Irene's bands some containing values larger than 100 mm h^{-1}. These are instantaneous measurements and over time accumulations caused significant flooding which brought 43 deaths and ~ 20 billion \$ damage to the NE coast of the USA. The obviously large spatial extent of Irene amply justifies use of surveillance scan for maximum storm coverage and Doppler scan for wind hazard detection.

Atmospheric biota is routinely observed with the WSR-88D network (Rinehart, 2010). Examples are insects, birds, and bats. Many insects are passive wind tracers providing a way to estimate winds in the planetary boundary layer (extending up to 2 km above ground).

Biota can be tracked for ecological or other purposes. The radar can also provide location of bird migrating paths, roosts, and other congregating places; this could be important for aircraft safety. The three donut shaped features in Fig. 19 represent Doppler speeds of birds leaving roost early in the morning. The critters are diverging away from the roost in search of food. Close to the radar the continuous field of velocities is principally from reflections off insects filling a good part of the boundary layer (this is deduced from polarimetric signatures, but not shown here).

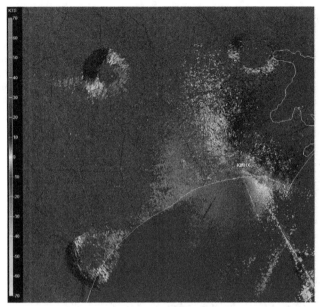

Fig. 19. Field of velocities obtained from the radar at Moorhead City, NC, on July 27, 2011 at 5:08 in the morning. The color bar indicates categories in kts; red away from the radar and green is toward. Elevation is 0.5°.

6. Epilogue

The WSR-88D network has been indispensable for issuing warnings of precipitation and wind related hazards in the USA. And its real time display of storm locations has become one of most popular and common applications on cellular phones. Its role in quantitative precipitation estimation is matching that of rain gages. So, what is beyond these achievements for the WSR-88D? Dual polarization upgrade combined with Doppler capability is the panacea a radar with the dish antenna on a rotating pedestal can achieve. Promising possibilities are: polarimetric confirmation of tornado touchdown at places where Doppler velocities indicate rotation; improvement of ground clutter filtering; polarimetric spectral analysis for extracting/separating features within radar resolution volume; significant improvement in data interpretation; inclusion of wind and precipitation type/amount in numerical prediction models; and other. Clearly the evolutionary trend continues and will do so at a decelerating pace until a plateau is reached. Complementary shorter wavelength (3 cm and 5 cm) surveillance radars are being considered for closing gaps or providing extra coverage at opportune places. (The TDWRs 5 cm wavelength radar data has been supplied to the NWS for several years). Explored are networks of tightly coordinated 3 cm wavelength radars for surveillance close to the ground.

One emerging technology is rapid scan agile beam phased array radar. This might be the ultimate radar providing it exceeds all the capabilities on the current network at faster scan rates. If in addition it proves to fulfill security and aviation needs (tracking of airplanes, missiles) it could revolutionize the current radar paradigm.

7. Acknowledgment

The author is grateful to Rich Ice, Darcy Saxion, Alan Free, and Dave Zittel for advice and valuable information about the WSR-88D. Sebastian Torres provided several figures and comments concerning technical aspects and designed signal processing for the MPAR. Dave Warde contributed figures and details about ground clutter and some VCPs. Collaboration with Dick Doviak is reflected in the requirements section. Allen Zahrai was in charge of engineering developments on MPAR and KOUN; Doug Forsyth lead the MPAR team in outstanding support and development of that platform.

8. References

Bachmann, S., & D. Zrnic (2007). Spectral Density of Polarimetric Variables Separating Biological Scatterers in the VAD Display. *J. Atmos. Oceanic Technol.*, Vol. 24, pp.1186–1198.

Bringi, V. N., & V. Chandrasekar (2001). *Polarimetric Doppler Weather Radar*. Cambridge University Press, Cambridge, UK.

Brown, R. A., V. T. Wood, & D. Sirmans (2002). Improved tornado detection using simulated and actual WSR-88D data with enhanced resolution. *J. Atmos. Oceanic Technol.*, Vol. 19, pp. 1759-1771.

Brown, R. A., B. A. Flickinger, E. Forren, D. M. Schultz, D. Sirmans, P. L. Spencer, V. T. Wood, & C. L. Ziegler (2005). Improved detection of severe storms using experimental fine-resolution WSR-88D measurements. *Weather and Forecasting*, Vol. 20, 3-14.

Burgess, D. W., V. T. Wood, & R. A. Brown (1982). Mesocyclone evolution statistics, *Severe Storm Conf. Proc.*, pp. 422-424, AMS, Boston, MA, USA.

Chrisman, J.N. (2009). Automated volume scan evaluation and termination (AVSET). *34th Conference on Radar Meteorology*, AMS, Williamsburg, VA, USA.

Crum, T.D., & R.L. Alberty (1993). The WSR-88D & the WSR-88D operational support facility." *B. American Meteorological Society*, Vol. 74, pp. 1669-1687.

Crum, T.D., R.L. Alberty, & D.W. Burgess (1993). Recording, archiving, and using WSR-88D data." *B. American Meteorological Society*, Vol. 74, pp. 645-653.

Curtis, D.C., & S. M. Torres (2011). Adaptive range oversampling to achieve faster scanning and the national weather radar testbed phased array radar. *J. Atmos. Oceanic Technol.*, in press.

Doviak, R.J., & D. S. Zrnic (2006). *Doppler radar and weather observations*. Second edition, reprinted by Dover, Mineola, NY, USA.

Free, A.D., & N.K. Patel (2007). Clutter censoring theory and application for the WSR-88D. *32nd Conference on Radar Meteorology*, AMS, Albuquerque, NM, USA.

Heiss, W.H, D.L McGrew, & D. Sirmans (1990). Next generation weather radar (WSR- 88D). *Microwave J.*, Vol. 33, pp. 79-98.

Hubbert, J.C., M. Dixon, & S.M. Ellis (2009). Weather radar ground clutter. Part II) Real- time identification and filtering. *Jour. Atmosph. Oceanic. Tech.* Vol. 26, pp. 1181-1197.

Ice, L.R., & D.S. Saxion (2011). Enhancing the foundational data from the WSR-88D) Part II, the future, *35th Conference on Radar Meteorology*, AMS, Pittsburgh, PA, USA. Ivic, R.I., D.S. Zrnic, & S. M. Torres (2003). Whitening in range to improve weather radar spectral moment estimates. Part II) Experimental evaluation. *J. Atmos. Oceanic Technol.*, Vol. 20, pp. 1449-1459.

Mahapatra, P. (1999). *Aviation weather surveillance systems.* Published by IEE and AIAA, printed by Short Run Press, Ltd, Exeter, UK.

Meischner, P. (2004). *Weather radar, principles and advanced applications.* Springer-Verlag, Berlin, Germany.

Meymaris, G., J.K. Williams, & J.C. Hubbert (2009). Performance of a proposed hybrid spectrum width estimator for the NEXRAD ORDA. *25th Int. Conf. on IIPS.* AMS, Phoenix, AZ, USA.

McLaughlin, D., & Coauthors (2009). Short-wavelength technology and the potential for distributed networks of small radar systems. *Bull. Amer. Meteor. Soc.*, Vol. 90, pp. 1797–1817.

Moisseev, D.N., & V. Chandrasekar (2009). Polarimetric spectral filter for adaptive clutter and noise suppression. *J. Atmos. Oceanic Technol.*, Vol. 26, 215-228.

Rinehart , R.E. (2010). *Radar for meteorologists.* Fifth edition. Rinehart Publications, Nevada, MO, USA.

Sachidananda, M., & D.S. Zrnic (2003). Unambiguous Range Extension by Overlay Resolution in Staggered PRT Technique. *J. Atmos. Oceanic Technol.*, Vol. 20, pp. 673-684.

Sachidananda, M., & D.S. Zrnic (1999). Systematic phase codes for resolving range overlaid signals in a Doppler weather radar. *J. Atmos. Oceanic Technol.*, Vol. 16, pp. 1351-1363.

Saxion, D, S., & R. L. Ice (2011). Enhancing the foundational data from the WSR- 88D) Part I, a history of success. *35th Conference on Radar Meteorology,* AMS, Pittsburgh, PA.

Serafin, R.J. & J.W. Wilson (2000). Operational weather radar in the United States Progress and opportunity. *Bull. Amer. Meteor. Soc.*, Vol. 81, pp. 501-518.

Siggia, A.D., & R. E. Passarelli, Jr. (2004). Gaussian model adaptive processing (GMAP) for improved ground clutter cancellation and moment calculation. *Proceedings of ERAD (2004)*, pp. 67-73. Visby, Island of Gotland, Sweden.

Torres, M.S., C.D. Curtis, D.S. Zrnic, & M. Jain (2007). Analysis of new Nexrad spectrum width estimator. *33rd Inter. Conf. on Radar Meteorology*, AMS, Cairns, Australia.

Torres, M.S., Y.F. Dubel, & D.S. Zrnic (2004a). Design, implementation, and demonstration of a staggered PRT algorithm for the WSR-88D. *J. Atmos. Oceanic Technol.*, 21, 1389-1399.

Torres, M.S., C.D. Curtis, & J.R. Cruz (2004b). Pseudowhitening of weather radar signals to improve spectral moment and polarimetric variable estimates at low signal-to-noise ratios. *IEEE Trans. Geosc. Remote Sens.* Vol. 42, pp. 941-949.

Torres S., Sachidananda, M, & D. Zrnic (2004c). Signal Design and Processing Techniques for WSR-88D Ambiguity Resolution) Phase coding and staggered PRT, implementation, data collection, and processing. NOAA/NSSL Report, Part 8, available from http://publications.nssl.noaa.gov/wsr88d_reports/.

Torres S., D. Zrnic, & Y. Dubel (2003). Signal Design and Processing Techniques for WSR-88D Ambiguity Resolution) Phase coding and staggered PRT, implementation, data collection, and processing. NOAA/NSSL Report, Part 7, available from http://publications.nssl.noaa.gov/wsr88d_reports/.

Torres, S.M., & D.S. Zrnic (2003). Whitening in range to improve weather radar spectral moment estimates. Part I) Formulation and simulation. *J. Atmos. Oceanic Technol.*, Vol. 20, pp. 1443-1448.

Unal, C. (2009). Spectral polarimetric radar clutter suppression to enhance atmospheric echoes. *J. Atmos. Oceanic Technol.*, Vol. 26, pp. 1781-1797.

Warde, A. D., & S. M. Torres (2009). Automatic detection and removal of ground clutter contamination on weather radars. *34th Conference on Radar Meteorology*, AMS, Williamsburg, VA, USA.

Weber, M., J.Y.N. Cho, J.S. Flavin, J. M. Herd, W. Benner, & G. Torok (2007). The next generation multi-mission U.S. surveillance radar network. *Bull. Amer. Meteorol. Soc.*, Vol. 88, pp. 1739-1751.

Wood, V.T, R. A. Brown, & D. Sirmans (2001). Technique for improving detection of WSR-88D mesocyclone signatures by increasing angular sampling. *Weather and Forecasting*. Vol. 16, pp. 177-184.

Zittel, W.D. , D. Saxion, R. Rhoton, & D.C. Crauder (2008). Combined WSR-88D technique to reduce range aliasing using phase coding and multiple Doppler scans. *24th IIPS Conference*, AMS. New Orleans.

Zrnic, D. S., J. F. Kimpel, D. F. Forsyth, A. Shapiro, G. Crain, R. Ferek, J. Heimmer, W. Benner, T. J. McNellis, & R. J. Vogt (2007). Agile beam phased array radar for weather observations. *Bull. Amer. Meteorol. Soc.*, Vol. 88, pp. 1753-1766.

Zrnic, D., S. V. M. Melnikov, & I. Ivic, 2008: Processing to obtain polarimetric variables on the ORDA (final version) NOAA/NSSL Report available from http://publications.nssl.noaa.gov/wsr88d_reports/.

2

Automated Processing of Doppler Radar Data for Severe Weather Warnings

Paul Joe[1], Sandy Dance[2], Valliappa Lakshmanan[3], Dirk Heizenreder [4],
Paul James[4], Peter Lang[4], Thomas Hengstebeck[4], Yerong Feng[5], P.W. Li[6],
Hon-Yin Yeung[6], Osamu Suzuki[7], Keiji Doi[7] and Jianhua Dai[8]

[1]*Environment Canada*
[2]*Bureau of Meteorology,*
[3]*CIMMS/OU/National Severe Storms Laboratory,*
[4]*Deutcher Wetterdienst,*
[5]*Guandong Meteorological Bureau, China Meteorological Agency,*
[6]*Hong Kong Observatory,*
[7]*Japan Meteorological Agency,*
[8]*Shanghai Meteorological Bureau, China Meteorological Agency,*
[1]*Canada*
[2]*Australia*
[3]*USA*
[4]*Germany*
[7]*Japan*
[5,6,8]*China*

1. Introduction

Radar is the only operational tool that provides observations of severe weather producing thunderstorms on a fine enough temporal or spatial resolution (minutes and kilometers) that enables warnings of severe weather. It can provide a three- dimensional view about every five to ten minutes at a spatial resolution of the order of 1 km or less. The development and evolution of intense convective precipitation is closely linked to thunderstorms and so understanding of the microphysics and dynamics of precipitation is needed to understand the evolution of thunderstorms as diabatic and precipitation processes modify and create hazardous rain, hail, wind and lightning.

The characteristics and proportion of severe weather is climatologically or geographically dependent. For example, the highest incidence of tornadoes is in the central U.S. whereas the tallest thunderstorms are found in Argentina (Zipser et al, 2006). Warning services developed at National Hydrological and Meteorological Services (NHMS) often originate because of a particular damaging severe weather event and ensuing expectations of the public. Office organization, resources and expertise are critical considerations in the use of radar for the preparation of severe weather warnings. Warnings also imply a level of legal liability requiring the authority of an operational National Hydrological Meteorological

Service. All available data and timely access is critical and requires substantial infrastructure, ongoing support and maintenance. Besides meteorological data, eye witness observations and reports are also essential element in the issuance of tornado warnings (Doswell et al, 1999; Moller, 1978).

This contribution will discuss operational or operational prototypical radar processing, visualization systems for the production of convective severe weather warnings. The focus will be on the severe weather identification algorithms, the underlying philosophy for its usage, the level of expertise required, decision-making and the preparation of the warning. Radar is also used for the precipitation estimation and its application for flash flood warnings. This is discussed elsewhere (Wilson and Brandes, 1979). Only a few countries have convective thunderstorm warning services and the target audience for this contribution are those countries or NHMS' considering developing such a service. The intent is to provide a broad overview and global survey of radar processing systems for the provision of severe weather warning services. There is a considerable literature in convective weather forecasting and warning, this contribution can only explore a few aspects of this topic (Doswell, 1982: Doswell, 1985; Johns and Doswell, 1992; Wilson et al, 1998).

The forecasting and the warning of severe weather are very briefly described. Then, the underlying technique for the identification of severe thunderstorms using radar is presented. This forms the basis for the radar algorithms that identify the severe storm features. The basic components of the system are then described. Some details and unique innovations are incorporated in the global survey of operational or near operational use. This is concluded by a summary.

2. Forecasting/Nowcasting/Severe weather warnings

Severe weather predictions are divided into severe weather watches and severe weather warnings. In the preceding days, thunderstorm outlooks may be issued. Watches are predictions of the potential of severe weather. They are strategic in nature and fairly coarse in spatial and temporal resolution. They are often issued on a schedule or in conjunction with the public forecast. The expected behaviour is that the public would be aware of the possibility of severe weather and to listen for future updates. Warnings are predictions of the occurrence or imminent occurrence (with high certainty) of severe weather. They are tactical and more specific in location and time. They are also specific in weather element. They are a call to action and to protect one's property and one's self. They are issued and updated as necessary. Fig. 1 shows an overview of the process from the Japanese Meteorological Agency.

Weather advisories are issued if the weather is a concern but not hazardous. Specific types of warning, such as tornado or hail warnings may then be issued and generally after the more generic severe thunderstorm warning is issued.

The key difference is that the watch is a forecast or very short range forecast service as strategic in nature whereas the warning is a nowcast (based on existing data, precise in time, location and weather element) and tactical in nature.

2.1 Severe weather definition

Severe weather is defined here as heavy rains, hail, strong winds including tornadoes and lightning. In the production of warnings, thresholds need to be defined. The thresholds are necessarily locally defined by climatology, local infrastructure and familiarity will dictate what is extreme. Table 1-4 show the warning criteria for Canada circa 1995. Canada is a very big country covering many different weather climatologies and therefore is illustrative of the variation of the severe weather thresholds (see also Galway, 1989). For example, Newfoundland on the east coast of Canada is a very windy location and hence strong winds are a common occurrence and the people have adapted to their environment and therefore it has the highest wind threshold in Canada. Each service needs to define these for them selves.

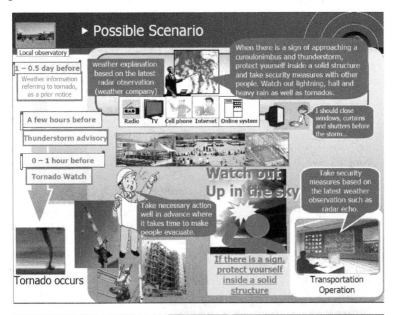

Fig. 1. The envisioned warning process from outlook to tornado watch. This is typical of the process that is used in most countries providing severe weather warning services. Getting the message out to and understood by the public is very important aspect of the utility of the warning service. Superimposing the warning on television, internet, mobile devices and directed messaging are critical to have the message heard.

Type	Description
Wind	Strong winds that cause mobility problems and possible damage to vegetation and structures.
Heavy Rainfall	Heavy or prolonged rainfall accumulating on a scale sufficient to cause local/widespread flooding.
Thunderstorm	One or more of the following: strong winds causing mobility difficulty, damage to structures due to wind and hail, heavy rain that may cause local flooding and lightning
Severe Weather	Presence of tornado(es), damaging hail, heavy rain, strong winds, life and property exposed to real threat, lightning
Tornado	Public has real potential to be exposed to tornado(es).

Table 1. Severe Weather Criteria in Canada: Warning Elements

Weather Centre	Wind	Rain	Hail	Remarks
Newfoundland	gusts of 90 km/h	25 mm/h	20 mm	No tornado criteria; no tornado warning; may mention hurricanes in marine warning.
Maritimes	gusts of 90 km/h	25 mm in 1 hr or 50 mm in 3 hrs	15 mm	Tornado or tornadic waterspout; no tornado warning; will issue hurricane prognostic message and information statements.
Quebec	gusts of 90 km/h	25 mm in 1 hr or 50 mm in 12 hrs	20 mm	Tornado, water spout, funnel cloud, windfall; no tornado warning
Ontario	gusts of 90 km/h	50 mm/hr for 1 hr; 75 mm for 3 hrs	20 mm	No tornado criteria in severe thunderstorm warning; tornado watch issued when confirmed tornadoes threaten to move into region or issued up to 6 hours in advance based on analysis or immediately for severe thunderstorms that indicate potential for becoming tornadic; tornado warning on forecast or observation.
Prairie	90 km/h	50 mm in 1 hr; 75 mm in 3 hr	20 mm	Tornado or waterspout probable; tornado warning issued when expected or observed.
Alberta	gust of 90 km/h	30 mm/h	20 mm	Tornado, waterspout or tornado warning when observed or expected or waterspout exists; cold air funnel cloud warning when cold air funnels expected but not tornadoes.
Arctic	90 km/h	25 mm/hr	12 mm	Tornado, water spout, funnel cloud; tornado occurrence warning on a confirmed report.
Yukon	gust to 90 km/h	25 mm in 2 hr	significant hail	Potential of tornado; warning is for thunderstorms; no tornado warning.
Pacific	gusts of 90 km/h	25 mm in 1 hr	15 mm	Lightning intensity of 500 strikes in 1 hr over an area of 1 degree x 1 degree latitude/longitude; no watches for severe thunderstorms or tornadoes; no tornado warnings; thunderstorm warning issued on a less than severe thunderstorm; will issue hurricane prognostic messages and information statements.

Table 2. Severe Weather Criteria in Canada: Severe Thunderstorm Criteria

Weather Centre	Warning Criteria
Newfoundland	50 mm in 24 hrs
Maritimes	50 mm in 24 hrs
Quebec	50 mm in 24 hrs or 30 mm in 12 hrs during a spring thaw
Ontario	50 mm in 12 hrs; sodden ground/bare frozen ground: 25 mm in 24 hrs; spring: 25 mm in 24 hrs; slow moving thunderstorms: 50 mm/3 hrs or 25 mm/3 hrs if ground is sodden.
Prairie	80 mm in 24 hrs or 50 mm in 12 hrs
Alberta	50 mm in 24 hrs
Arctic	50 mm in 24 hr
Yukon	40 mm in 24 hr
Pacific	50 mm in 24 hr except in west Vancouver Island and northern coastal regions 100 mm in 24 hr and interior of B.C. 25 mm in 24 hr

Table 3. Severe Weather Criteria in Canada: Heavy Rainfall Warning

Weather Centre	Warning Criteria
Newfoundland	75 km/h and/or gusts of 100 km/h
Maritimes	65 km/h and/or gusts to 90 km/h
Quebec	50 km/h with gusts to 90 km/h or with only gust to 90 km/h
Ontario	60 km/h for 3 hours, or gusts of 90 km/h for 3 hrs
Prairie	60 km/h and/or gusts to 90 km/h for 1 hr
Alberta	60 km/h or gusts to 100 km/h except in Lethbridge Region: 70 km/h or gusts to 120 km/h.
Arctic	60 km/h or gusts of 90 km/h
Yukon	60 km/h for 3 hr or gusts to 90 km/h
Pacific	Mandatory 90 km/h expected over adjacent marine areas; discretionary if gale force winds (63 to 89 km/h) expected over marine areas; discretionary for interior B.C. 65 km/h or gusts of 90 km/h

Table 4. Severe Weather Criteria in Canada: Strong Wind Warning

Warnings for summer severe weather are for extreme or rare events - events that are at the high end of the spectrum of weather. In terms of statistics, rare events do not occur very often (by definition) and so statistical analyses are always suspect due to low numbers. It is difficult to easily demonstrate (using statistics) the efficacy of a warning program (Doswell et al, 1990; Ebert et al 2004). Qualitative analyses or case studies are required to understand the relationship between the provision of warnings and the saving of lives (Sills et al, 2004; Fox et al, 2004). The same applies to determining the efficacy of radar algorithms to the provision of weather warnings (Joe et al, 2004).

This has a significant impact on statistics but also on the "cry wolf" syndrome (AMS, 2001; Barnes et al, 2007; Schumacher et al, 2010; Westefeld et al, 2006). An accurate but useless tornado forecast could be by stating that "next year there will be a tornado in the U.S." This statement is a climatological or statistical forecast. It has a very high probability of being true. However, the phenomenon is very small, perhaps 10-20 km in length and 500 m in width and so this particular prediction is not very useful. The information is highly accurate but not very precise in terms of location or time. Most, if not all, people would ignore the warning and take the risk. Another form of the "cry wolf" syndrome is where warnings are issued indiscriminately for a very precise time and location and with considerable lead time. However, particularly for rare events (those at the extreme end of a distribution), this is accompanied by a high false alarm rate. If too many false alarms are issued, then these will also be ignored. So, for rare extreme hazardous events, high probability of detection is needed but the false alarms need to be mitigated (Bieringer and Ray, 1996; Black and Ashley, 2011; Glahn 2005; Hoekstra et al, 2011; Polger et al, 1994).

So the issuance of warnings requires a very fine balance of decision-making that takes into account lead time, climatology, societal risk behaviour, social-economic infrastructure, warning service capacity and many other regional, political and societal factors (Baumgart et al, 2008; Dunn, 1990; Hammer and Schmidlin, 2002; Mercer et al, 2009; Schmeits et al, 2008; Westefeld et al, 2006; Wilson et al, 2004). Nowcasts in general are user dependent (Baumgart et al, 2008). Warnings are an extreme kind of nowcasts in which the thresholds apply to a very broad range of users (the public). However, in the future, one can envision very specific warnings or nowcasts issued at lower thresholds that may affect specific users requiring tailored communication techniques and technologies (Keenan et al, 2004; Schumacher et al, 2010).

The wind hazard deserves an extended discussion (Doswell, 2001). There are various kinds of wind hazards that have distinctive life times and spatial features. Straight line winds can originate in synoptic systems or typhoons and are ubiquitous, broad in spatial scale (~100+ km) and extended in duration (~hours/days). Derechos[1] are also straight line winds that originate out of mesoscale convective complexes (MCC; Davis et al, 2004; Evans and Doswell, 2001; Przybylinski, 1995; Weisman, 2001). The damaging portion exists at specific locations. They are smaller in size and temporal scale than the previous kind of winds. Gust fronts originate with the downdrafts of MCC's and depending on the nature of the MCC (isolated thunderstorm, multi- cellular, line echo wave pattern, bow echo, pulse storm); the gust front can take on many forms but generally emanate outwards from the MCC (Klingle et al, 1987). They can extend for a long time and there may be extreme winds in portions of the gust front.

The downdrafts can also generate quasi-circular outward flowing winds called downbursts (generic term). If the downbursts are over airports, small in diameter (<4km) and intense (>10 m/s velocity differential) then they are given a very specific term called the microburst (McCarthy et al, 1982; Wilson et al, 1988; Wilson and Wakimoto 2001). It is arbitrarily defined this way in order to be very clear to aviators that they are hazardous and should not be transected. They originate with a descending intense precipitation core and the wind intensity is enhanced by evaporative cooling (Byko et al, 2009). If evaporation is strong, by the time the downburst reaches the surface, there may not be any precipitation associated with it. In this case, the feature is called a dry downburst. If there is precipitation then it is called a wet downburst or microburst as the case may be.

There are algorithmic radar techniques for the identification of all of these severe weather features (Dance and Potts, 2002; Donaldson and Desrochers, 1990; Johnson et al, 1998; Joe et

---

[1]It is beyond the scope of this contribution to illustrate the various severe hazards in detail – see references for fourther information.

al, 2004; Kessler and Wilson, 1971; Lakshmanan et al, 2003; Lakshmanan and Smith, 2009; Lakshmanan et al, 2009; Lenning et al, 1998; Mitchell et al, 1998; Stumpf et al, 1998; Winston, 1998; Witt et al, 1998a, Witt et al, 1998b). The efficacy of the detection depends on the radar scan strategy and quality of the radar (range, azimuth resolution, cycle time, sensitivity, elevation angles, number of elevation tilts, etc (Brown et al, 2000; Heinselman et al, 2008; Lakshmanan et al, 2006; Marshall and Ballantyne, 1975; McLaughlin et al, 2009; Vasiloff, 2001).

2.2 Watches

Watches are based on the concept that the juxtaposition of dynamics, thermodynamics and a mechanism to create upward motion and/or a mechanism to remove inhibition factors exists. This is often called the ingredients approach as one looks to see where the various ingredients come together and that is where severe weather will occur. Historically, this is based on the original Fawbush and Miller Technique (1953) but it has gone through significant evolution (Doswell, 1980, 1982, 1985, 2001; Johns and Doswell, 1992; Moller, 2001; Moninger et al, 1991; Monteverdi et al, 2003; Rasmussen, 2003; Weiss et al 1980).

Fig. 2 shows the morphology of thunderstorms that theoretically develop under different wind shear and convective available potential energy (CAPE) situations (Brooks et al, 1993; Brooks et al, 1994; Markowski et al, 1998b; Weisman and Klemp, 1984; Weisman and Rotunno, 2000). Dynamics is represented by the 0-3 km magnitude of the wind shear. Other height limits may be used depending on the region and local operational usage. The atmospheric structure (low level moisture, mid level dry air, strength of inversions, etc) is important and the thermodynamics is represented by CAPE in this figure. While shear and CAPE are two basic indices that are often used, many other indices are investigated and used.

length of 0-6 km shear vector (kt)

	<20	20-45	>40
CAPE (J/kg) <1000	ORDINARY	ORDINARY OR MULTICELL	ORDINARY OR SUPERCELL
1000-2500	ORDINARY WITH SOME PULSE SEVERE	MULTICELL	SUPERCELL
>2500	ORDINARY WITH SOME PULSE SEVERE	MULTICELL	SUPERCELL

Fig. 2. Thunderstorm type as a function of CAPE and Shear.

Watches are generally very broad in spatial nature due to the spatial density of the observations (soundings and surface observation), and models which are based on the observations, which is very sparse. The resulting analysis of severe weather potential is therefore necessarily broad. The situation is also very fluid and there can be many local factors such a topography or land-water boundaries or rural- urban differences, to name just a few (King et al, 2003; Wasula et al, 2002; Wilson et al, 2010). What are very difficult to identify are potential mechanisms to create upward motion (the trigger) or to overcome the convective inhibition (break the cap). On a synoptic scale, this could be lift generated by cold or warm fronts but on a smaller scale, they can be created by dry lines, thunderstorm outflows, lake-land breezes, urban hot spots, etc. Often they are very low level and therefore hard to observe. So forecasts of severe weather are indications that the potential ingredients exist. They are therefore very broad and strategic in nature.

2.3 Warnings

Weather warnings are issued when there is very high likelihood of severe weather. A broadly worded severe weather thunderstorm warning is most often first issued. If appropriate, it is followed by a more specific warning on a particular thunderstorm and specific severe weather element. This approach is not universal but is dependent on the climatology of severe weather and the level of the warning service that can or has been decided to provide. An important aspect of the detail of the warning is the ability to use the information by the end-user, which is often the public. The public may not know how to react. Given the "cry wolf" syndrome, there needs to be an education process (see Fig. 1). Often, a disaster is needed to get the attention of the public but the significance of the event can be lost in a few short years. Civil emergency services and hydro utilities can plan their post- event remediation actions/locations based on the warning areas and products. So, there can be many variations and underlying philosophies for the provision of warning services. This partially drives the design of the radar processing, visualization and warning preparations systems. It is one thing if severe weather is prevalent and there is a dedicated forecaster for a small area and the public is well attuned to the severity of the weather and have tornado shelters (Andra et al, 2002). It is another thing if the forecaster has to cover several radars and dealing with ill informed users (Leduc et al, 2002; Schumacher et al, 2010).

3. Identifying severe thunderstorms

3.1 Lemon technique

The specificity of the severe thunderstorm warning is primarily based on a radar feature identification technique attributed to Lemon (1977, 1980) and is based on a morphological approach (Moller et al, 1994). It is beyond the scope of this contribution to present or describe the various types of thunderstorms (Fig. 3 shows a small sample). As mentioned earlier, precipitation and precipitation cores form aloft and then descend.

The following features need to be identified:

- tilted updraft, and/or weak or bound weak echo region
- displaced echo top relative to the low-mid level core
- strong reflectivity gradients

- high low level reflectivity core displace towards the updraft
- concavity (hook echo)
- deviant motion (right or left mover, depending on hemisphere)
- rotation

This is a highly condensed version of the technique and there are many subtleties and morphological pathways as storms evolve. Severe storms begin as non-severe storms and algorithm developers and forecasters try very hard to extend lead times by trying to identify the severity of the future storm as early as possible. Note also that it is often in the collapsing stages of the storm (indicated by collapsing echo top or a descending core) when the severe weather reaches the surface (see Fig. 4).

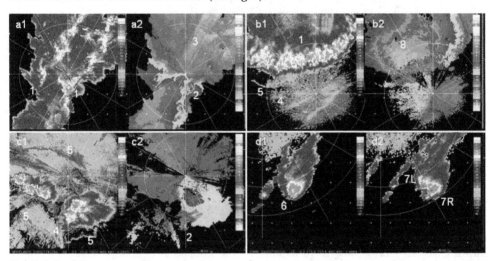

Fig. 3. It is obligatory to show radar images of severe convective storms. Linear convective storms are show in (a) and (b) whereas isolated thunderstorms are shown in (c) and (d). Except for (d), reflectivity and radial velocity images are shown together. Fig. 3a shows double squall lines (1) with embedded cells and mesocyclones (2). (3) shows a shear line associated with a cold frontal passage, so the mesocyclones are pre-frontal and likely to have formed on a previously formed outflow boundary. Fig 3b shows embedded thunderstorms on a bow echo. Note the boundaries (5) ahead of the bow echo. (8) shows a meso-scale intense straight line wind (nearing 48 m/s). Fig 3c show an isolated thunderstorm with a mesocyclone (4). Boundaries (5) can be seen and to be associated with the entire mesoscale convective complex and not just one individual cell. Fig. 3d shows the splitting of an isolated tornado producing storm. The yellow shading is the 40 dBZ contour. Often, cell identification thresholds are set lower (30 or 35 dBZ) in an attempt to get earlier cell detections but this demonstrates that this results in detecting different storm structures.

Not discussed here is the identification of the initiation phase of convective weather (Wilson et al, 1998). Significant progress has been made in the warning of air mass thunderstorms. In the past, these were considered random and unforecastable. Wilson et al (1998) demonstrate that they are not random but form on boundaries (see the fine lines on Fig. 3c). Roberts et al

(2006) discuss the tools to help bridge the convective initiation phase to the severe phase of thunderstorm nowcasting. The science or theory of thunderstorm is still evolving (Brooks et al, 1994; Brunner et al, 2007; Markowski, 2002; Rasmussen et al, 1994; Weisman and Rotunno, 2004).

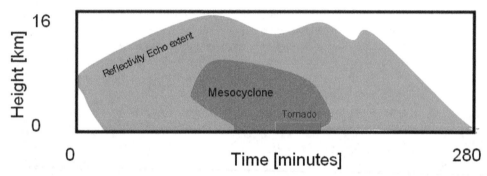

Fig. 4. A time-height diagram through the core of a long lived thunderstorm with a mesocyclone. The "nose" on the left side of the shading indicates the precipitation and the mesocyclone originate at mid-levels of the atmosphere and develop vertically up and down. In the collapse phase of the storm or mesocyclone top, the severe weather reaches the ground (adapted from Burgess et al, 1993; Lemon and Doswell, 1979).

3.2 Other data sets

This contribution focuses on radar and its use in the preparation of warnings. In fact, all sources of observations and information are used to validate and enforce the conceptual models used to produce the warnings. Satellite imagery, such as provided by MSG and the future GOES-R, will be able to provide 5 minute updates over limited areas. Lightning networks are now prevalent and often used as surrogates for radar data where none is available. They also directly observe the lightning hazard (Branick et al, 1992; Gatlin et al, 2010; Goodman et al, 1988; Knupp et al, 2003; Lang et al, 2004; Schultz et al, 2011;). Even though a single lightning flash can cause serious harm or death, table 2 indicates that, in Canada, a propensity of lightning strikes is needed before a lightning warning will be issued. Surface wind reports can be also used. However, a tornado or a microburst is relatively small and most operational networks are too sparse to effectively sample the atmosphere for such a small feature. At some airports, a dense network of anemometers is established for this specific problem (Wilson et al, 1998). An important data set are eye witness reports (Doswell et al, 1999; Moller 1978; Smith, 1999). In the past, eye witness reports were required before a tornado warning would be issued. This made all tornado warnings "late" with negative lead times. This was done in order not to "cry wolf" and "alarm the public". An emerging source of information is the use of high resolution NWP (Hoekstra et al 2011; Li, 2010; Stensrud et al 2009). While phase errors exist (time and location of the thunderstorm), the models appear to be able to capture the morphology of the storm (see Fig. 2). While radar is the core observation system for severe weather warnings at the convective scale, these are not available everywhere. A warning service that does not include radar has yet to be effectively demonstrated.

3.3 Radar dependencies

The Lemon technique implies that volume scanning radars are needed since many of the critical features originate aloft (see Fig. 4). Both, high data quality (Joe, 2009; Lakshmanan et al, 2007; Lakshmanan et al 2010; Lakshmanan et al, 2011) and rapid update cycles for the fast evolving thunderstorms (Crum and Alberty, 1993; Marshall and Ballantyne, 1975). In order to detect low level "clear air" boundaries important for the identification of convective initiation, high sensitivity is critical. Research literature often shows many examples of extensive clear air radar echoes that are not operationally observed. The operational question is whether it is a radar sensitivity issue or the lack of insect targets (the clear air targets have been identified as insects through dual-polarization signatures). Extensive clear air echoes are commonly reported observed on the WSR-88D and primarily in certain parts of the United States (Wilson et al, 1998). Table 5 shows the sensitivity of a small sample of radars including the WSR-88D, WSR-98D (S Band radars) and three C Band radars, one of which is a low powered (8 kW), travelling wave tube (TWT) solid sate pulse compression radar (Joe, 2009; Bech et al, 2004; O'Hora and Bech, 2007). In units of dBZ, the sensitivity is a function of range. Fifty kilometer range is arbitrarily chosen to compare the radar sensitivities. The table shows that all these state of the art radars can have comparable sensitivity. Therefore, the apparent lack of clear air echoes is due to the lack of local clear air radar targets and not due to radar sensitivity or wavelength (for example, see May et al, 2004). In addition, due to the dependendence of the backscatter on the inverse frequency squared, C Band radars should observe insects better than S Band radars.

Radar	MDS at 50 km
WSR-98D (TJ)	-6.0dBZ
WSR-98D (BJ)	-5.5 dBZ
WSR-88D (KTLX)	-7.5 dBZ
WSR-88D (KLCH)	-8.5 dBZ
WKR Conventional C Band (2 µs pulse)	-11.0 dBZ
WKR Conventional C Band (0.5 µs pulse)	-5.0 dBZ
CDV TWT (8kW) C Band (1 µs pulse)	6.0 dBZ
CDV TWT (8kW) C Band (5 µs pulse)	-7.0 dBZ
CDV TWT (8kW) C Band (NLFM 30 µs pulse)	-6.0 dBZ
CDV TWT (8kW) C Band (NLFM 40 µs pulse)	-9.0 dBZ
INM Conventional C Band (2 µs pulse)	-9.0 dBZ

Table 5. Minimum Detectable Signal of Various Radars

4. Forecast process and system design

Perhaps the most important consideration in the design of the operational radar processing, visualization and decision-making is the underlying philosophy of the weather service, existing systems and, of course, the capabilities and resources available (Joe et al 2002). In many cases, the warning service requirements are driven not only by the scientific capabilities or the needs but also by the political, societal and economic norms. Often a warning service is an ethical and moral reaction by NHMS's to a damaging event or events and hence it is also a political reaction by governments. This varies considerably from place to place. These requirements are tempered by existing observational infrastructure. Are there functioning radars or other data sources? Is there the capacity to design or even adopt a radar processing system? Is there the knowledge and capacity to interpret the data products to make effective warning decisions and issue warnings? And is there a way to reach the end-user in a timely fashion? It should not be forgotten that the end-user must be educated on the meaning of the warning and on how to react appropriately. Is there sufficient budget to develop a warning system? What is risk is acceptable? What level is the moral outrage?

An often overlooked design issue is the organization of the weather service. Warnings are provided for small areas (scale of the weather feature) in order to mitigate the "cry wolf" syndrome to be effective (Barnes et al, 2007; Hammer and Schmidlin, 2002). The critical issue is the capacity to provide the attention to the detail given the totality of the forecast responsibilities. The system design will be quite different if there are many forecast offices and few radars (one to one) compared to few offices and many radars (one office to ten radars as in Canada).

Of course, an overarching issue is the climatology of severe weather which ultimately is the core issue. For many countries, convective weather may occur year round and some only for the summer season. In the latter case, a design question is to determine the use case for the shoulder season where severe weather may occur unexpectedly and the warning service is seasonal.

Severe weather forecasting requires a unique forecasting skill set. In synoptic forecasting (for 12 hours and beyond), the forecaster compares current observations to numerical weather prediction models to evaluate the appropriateness of the model or to develop a conceptual model of the weather for the creation of the public forecast product (Doswell 2004). The product is usually produced on a fixed schedule. In severe weather forecasting, the observations need to be timely; there is urgency in the interpretation and the generation of the warning product. It is a "short fused" situation. These require different personality types and this also drives the design considerations. In order to mitigate the "cry wolf" situation while maintaining high probability of detection, a dedicated and separate warning forecaster function is required to be able to address the immediacy issues of the warning service. These are just some of the design considerations for a radar processing and visualization system and the forecast process for the provision of severe weather warnings. Forecast process refers to all components of the transformation of the data or observations into information used for decision- making and warning service production. It includes both the human and their tools and is often referred to as the man-machine mix. Given all the degrees of freedom in the chain, there are different models of the forecast process.

In the next section, a global survey (necessarily incomplete) is presented that will briefly examine the operational or near-operational systems that have been developed. Many have commonalities and only the underlying unique aspects will be highlighted.

5. Components of a basic system

In this section, the basic components or issues of severe weather radar processing/visualization are briefly discussed and a block diagram is provide in Fig. 5. The benefits of different radar types are discussed elsewhere (WMO, 2008).

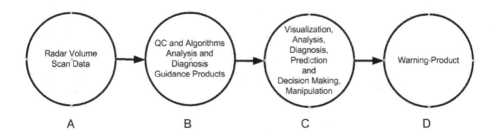

Fig. 5. The flow of the radar data to warning product is much the same in all systems. But the contents of each stage can be different. Except for one system described in this contribution, all the others require human decision-making at stage C before the warning product is issued to the public. In the case of KONRAD (see section 6.10), the product goes mainly to "sophisticated" users.

5.1 Data quality

Radar processing systems need quality controlled data. This can occur in a separate and independent process. In some cases, it is part of the adjustments and corrections that need to be made. Before the severe weather processing occurs (stage B in Fig. 5), it is assumed that the data is free of anomalous propagation, ground clutter and biases in power are adjusted. Second trip echoes and range folded may still be in the Doppler data (Joe 2009; Lakshmanan et al, 2010; Lakshmanan et al, 2011).

In high shear environments the assumption that the radial velocities within a range volume are uniform may not be satisfied (Holleman and Beekhuis 2003; Joe and May 2003). Fig 6ab shows a simulated Doppler velocity spectrum (based on an example in Doviak and Zrnic, 1984) of a tornado contained within a single range volume. The spectrum is bi-modal and the peaks at located at the speed of the radial components of the tornado. Normally it is uni-modal and Gaussian in shape. Fig. 6cd show the measured spectrum given two different Nyquist limits. The spectrum is aliased and overlaps with itself. The smaller the Nyquist limit, the greater the overlap. In highly sheared regions, the velocity data is noisy and can be non-sensical. The chapter on quantitative precipitation estimation addresses many of the quality control issues.

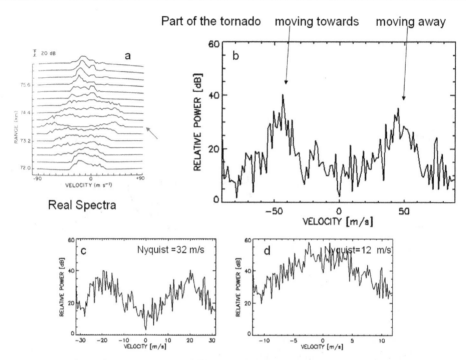

Fig. 6. (a) Doppler velocity spectra at different ranges made with a radar with a very large Nyquist interval. The arrow points to the tornado. The spectrum is bi-model. (b) a simulation of the spectra. (c) and (d) are simulated measured spectra made with different Nyquist intervals. The spectrum overlaps and is aliased. In (c) the spectra is bi-modal still, would produce a radial velocity estimate near zero with a very broad variance. In (d), the mean is still zero, the spectra is uni-modal with a smaller variance.

6. Global survey

This section provides a necessarily brief global survey of various convective weather radar processing systems. In fact, there are only a few NHMS' that actually provide a severe weather warning service. The systems are presented in a sequence that approximately matches when they were developed and the reader can follow the progression of the system and philosophical developments.

6.1 RADAP – II, U.S.A.

The first radar processing system for severe weather was RADAP-II and it was built in the 1970's (Winston and Ruthi, 1986) and it followed from D/RADEX (Breidenbach et al, 1995; Saffle, 1976) within the National Weather Service. They used VIL (vertically integrated liquid water) and a significant innovation was the introduction of a SWP (Severe Weather Probability) product. They were using probabilistic and uncertainty concepts then! There were many innovations with RADAP-II but its deployment was curtailed due to the development of the Doppler upgrade called the WSR-88D (Crum and Alberty, 1993; Lemon et al, 1977; Wilson et al, 1980). Crane (1979) developed the cell identification techniques based on peak detection. These systems left a legacy for the development of the WSR88D algorithms. McGill developed SHARP (Bellon and Austin, 1978) for precipitation nowcasting and developed the cross-correlation method for echo tracking which is still used today. It did not specifically address severe weather algorithms, which is the focus of this contribution.

6.2 WSR-88D, U.S.A., WSR-98D, China

Many of the innovations for the reflectivity-only algorithms of RADAP-II were adopted and significantly enhanced for the WSR88D (Crum and Alberty, 1993; Kitzmiller et al, 1995). Doppler algorithms were developed for mesocyclone and gust front detection (Hermes et al 1993; Uyeda and Zrnic, 1986; Zrnic et al, 1985). Considerable effort has been expended to improve upon these initial efforts. A search of the American Meteorological Society journal publications will illustrate that. Initially, the output from the WSR88D Radar Product Generator was displayed on a dedicated radar-only visualization system called the Principal User Product (PUP) display for the forecaster and later the forecaster workstation called AWIPS was used. This integrated all the data and products that the forecaster needed. WSR-88D algorithms were later deployed on the WSR-98D radars made by MetStar and used in China, Romania, India, Korea and other places.

A fundamental question arose as to the role of automated guidance products versus manual interpretation (Andra et al, 2002). It is clear that automated generated products are for guidance and it should not be mistakenly interpreted that warnings were automatically generated and issued without an intervening well trained decision-maker. Initially, there was an extensive radar training program for forecasters, up to 6 weeks for specialists. Clearly, the expectation was that an expert level of training was needed to interpret Doppler radar data for severe weather warnings. This was re-enforced by the work of Pliske et al (1997) who analyzed how to achieve the expected benefits of a modernization program. This resulted in the development of an on-going training program for decision-making at the appropriately named, Warning Decision Training Branch of the National Severe Storms Laboratory. Professionally trained instructors on cognitive principles interactively have the skills to tailor the material to the appropriate knowledge level, abilities and learning styles of the student. It is a model for professional training.

6.3 TITAN – NCAR

TITAN (Thunderstorm identification, tracking and nowcasting) was first developed in South Africa and then later at NCAR for support of weather modification programs. Dixon and Weiner (1993) described a simple but brilliant threshold technique for the identification of thunderstorm cell cores. This simplified the peak detection techniques of the Crane (1979) technique as the latter identified many weak cells and challenged the computing power of the day. It also described a methodology for tracking. It could be argued that this is the most widely used system in the world. It is freely available and requires some expertise to implement. It is used extensively in research environments (Lei et al, 2009). It is a stand alone system and integrating it into an operational environment has been done but there are capacity and support issues to consider. For example, it is used at the South African Weather Service.

6.4 WDSS-I and II – USA

WDSS-I was a research analysis tool and made great strides in developing innovative algorithms and concepts. WDSS-I (Eilts et al, 1996) processed single radar data. A particular innovation was the Storm Cell Identification and Tracking algorithm (Johnson et al, 1998) which ranked the storms by severity. This extended the SWP product from RADAP-II. This system is commercially available from Weather Decision Technologies. WDSS-II was an enhanced version of WDSS-I (Lakshmanan et al, 2006). It has a multi-radar capability and integrates other data. Fig. 7 shows a chart of the data processing flow and lists the algorithms. A technical innovation is in the handling of radar data in overlap regions. Radar cell identifications (and others such as mesocyclone detection) are first done along each PPI surface to identify 2D cell objects. Then these 2D objects are collated together into a 3D multi-radar object. A five minute window is used to aggregate the data and cells are time shifted to a common moment in time. A service innovation is that this extends the warning service capability to a regional level (more than the domain of single radar). WDSS-II saw the return to the display of more imagery to support experts in their decision-making (Fig. 8).

Fig. 8 shows shear fields and aggregated shear fields. While they were computed as part of the severe weather algorithms internal computations, they were not previously displayed. With the development of fast computers and display capabilities and the realization that expert forecasters can effectively use these products, they became in vogue.

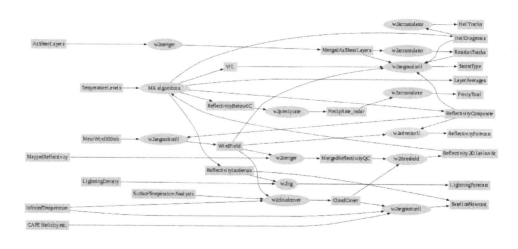

Fig. 7. The data flow of the WDSS-II system. This system integrates "other" data (numerical weather prediction data) including model data into the radar processing. While this is common for QPE applications to help identify the bright band or melting level, this was an innovation in severe weather processing.

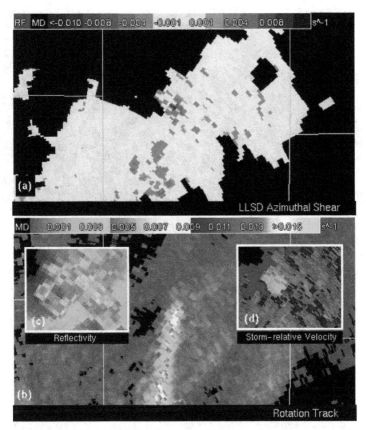

Fig. 8. In this system, there is a trend to go back to basic imagery products such as shear and aggregated shear to aid in the interpretation and utility of the data.

6.5 CARDS – Canada

The CARDS (Canadian Radar Decision Support) system was developed as part of the radar upgrade (Joe et al, 2002; Lapczak et al, 1999) and built on the previous concepts. In Canada, a single severe weather forecaster is responsible for the provision of warnings for the area coverage of about ten radars. This is in contrast to other countries, where it is approximately one radar for one forecaster. While this may seem like a work overload situation, there are some interesting side benefits. It has been estimated that in a one radar for one forecaster situation, a forecaster will likely face only one significant event in his career. In the Canadian scenario, a severe weather forecaster will therefore experience ten big events. It can be argued that these experienced forecasters will be better at decision making and will therefore make better warnings (Doswell, 2004). Forecasting is a complex process and it remains to be seen whether this is a true. Given these constraints, the weather service of Canada is arguably the most reliant on automated guidance products. They are critical in aiding the forecaster to diagnose those cells which need detailed interrogation to upgrade from a severe weather warning to a more specific warning.

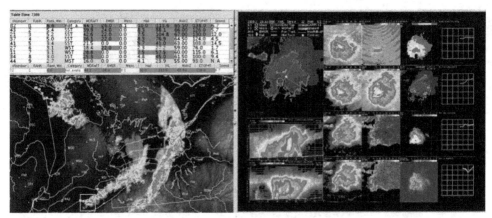

Fig. 9. An example of a CARDS composite, SCIT and cell view. The size of the forecast domain is about ~2000 km x 1600 km. The image shows a zoomed image of the cells, tracks and lightning strikes. Eight Canadian radars and 12 US radars contribute to the image. The composite and the SCIT table products are invoked and displayed at the same time. The forecaster can either drill down to a CELL VIEW via the composite or via the SCIT table. They can also rapidly survey the cells from the SCIT table without invoking the CELL VIEW products. The colour coding indicates the categorical ranking. On the right is an example of a cell view. This shows a variety of images that allows the forecaster to quickly make a decision as to the severity of the storm. The product shows an ensemble product of the algorithms (upper left hand corner, not described), automatically determined cross-sections, four CAPPIs (1.5, 3.0, 7.0, 9.0 km), reflectivity gradient, MAXR, echo top, VIL density, Hail, BWER and 45 dBZ echo top and time graphs.

In an envisioned future exercise for the design of CARDS, it was identified that there was actually no hard requirement for single radar products. One of the main reasons for missed warnings was that the forecaster was so intent on one thunderstorm that they forgot about the others. There was a loss of situational awareness. This happens even with experienced forecasters or analysts and is common in many fields where critical decisions are made. A regional composite that could display and overlay the most popular products (CAPPI, EchoTop, etc) is the main product to maintain situational awareness. Thunderstorms cell locations are identified, ranked, color coded and displayed on the composite and in a table similar to the SCIT table. Selecting the cell of interest in the composite or in the table, the user is able to quickly and rapidly drill down to reveal a cell view product (Fig. 9) that contain all the products that the user would use to interrogate a cell and make decisions. The cell view has a legacy from Chisholm and Renick (1972). The design exercise also identified the critical reliance on automated guidance products.

Another important innovation is that the visualization tool for the image and data products is based on hypertext transfer protocol (http) which means that any computer regardless of operating system can access the full functionality of the radar data. Analyzing breakthroughs in the use of radar, access to the data and the products has been "the" key innovation. Recall the days of radar operators who hand drew radar maps or the facsimile machine or the mono or color graphics terminal. Each innovation increased the capacity to deliver better products. In today's technology, every button press or mouse click that is eliminated delivers "a big bang for the buck". This key innovation allowed the data to be effectively used in the Sydney Olympic Command Centre (Joe et al, 2004; Keenan et al, 2004).

Similar to the SCIT of WDSS-I, CARDS implemented a fuzzy logic technique to rank storms. The technique is configurable (see Table 6). It shows the parameters that the users decided to use and the thresholds that they considered as weak, moderate, strong and severe (see also Doswell et al, 2006).

In the overlap region, cells are selected from one radar or the other, unlike WDSS-II. Due to attenuation concerns, lack of experience with the fuzzy logic storm severity technique and that reflectivity (and reflectivity based products) was still the prime parameter for determining storm severity; users selected the cell detection with the maximum reflectivity as the cell for visualization. However, this would likely not be the case anymore as nearest radar or maximum information or maximum severity ranking would be chosen today.

Thresholds	Rank (0-8)	BWER count	Meso m/s/km	Hail cm	Wdraft m/s	Vil density kg m^2/ km	Max Z dBZ	45 dBZ ETop km
	0	0	0	0	0	0	0	0
Minimum	1 0-2	5-11	4	0.5	10	2.2	30	5.5
Weak	2 3-4	12-17	6	1.3	15	3	45	8.5
Moderate	3 5-6	18-21	8	2.3	20	3.5	50	10.5
Severe	4 7+	22-26	10	5	25	4	60	12.5

Notes:

Rank:	5-6 means a value of 5 or more but less than 7.
WER:	The number of directions where reflectivity increases determines a BWER (with low reflectivity below)
Meso:	Average Pattern Vector Shear (see Zrnic et al, 1985)
Hail:	Average Hail Size
WDRAFT:	Gust potential in m/s
Vil Density:	Similar to WDRAFT in pattern
VIL	if VIL for classification then 10 20 30 40 are the thresholds
Max Z:	Max reflectivity in the cell
45 dBZ Echotop Ht:	Reliable echo top parameter

Table 6. Fuzzy Logic Membership Functions for Parameters Used to Rank Storms.

6.6 SWIRLS and its variants – Hong Kong, China

In Hong Kong, lightning strikes and damaging squalls are major threats accompanying thunderstorms. In support of the Thunderstorm Warning operations, SWIRLS (**S**hort-range **W**arning of **I**ntense **R**ainstorms in **L**ocalized **S**ystems) was developed to track and predict severe weather including rainstorms, cloud-to-ground (CG) lightning, damaging thunderstorm squalls and hail for the general public. The warning decision and message preparation are made by the Observatory's duty forecaster. Once issued, the warning message are disseminated automatically through various channels including radio and television broadcast automatic telephone enquiry system, Internet web page, as well as mobile apps for smart phones and social networking platforms such as Twitter.

An innovation is the DELITE (**D**etection of cloud **E**lectrification and **L**ightning based on **I**sothermal **T**hunderstorm **E**choes) algorithm for lightning warning. It selects radar and other parameters most relevant to the microphysical processes leading up to the electrification of a cumulus cloud (Fig. 10). This includes radar reflectivity at constant temperature levels (0°C, -10°C, and -20°C), the thermal profile of the troposphere (from either numerical weather model analysis or the latest available radiosonde data), the echo top height and the vertically integrated liquid (VIL). CG lightning initiation is expected if prescribed thresholds are exceeded.

The above severe weather analyses are performed on a cell basis and the threat areas are identified as elliptical cells in the corresponding interest fields with values greater than or equal to prescribed thresholds. For example, the detailed cell identification technique follows the GTrack algorithm of SWIRLS. For lightning and downburst, the interest fields are 3-km CAPPI and 0-5 km VIL respectively. The thresholds are 25 dBZ and 5 mm respectively.

MOVA (**M**ulti-scale **O**ptical flow by **V**ariational **A**nalysis) is a gridded echo-motion field that is derived from consecutive radar reflectivity fields by solving an optical-flow equation with a smoothness constraint. To capture multi-scale echo motions, the optical-flow equation is solved iteratively for a cascade of grids from coarse to fine resolutions (about 512 to 3 km).

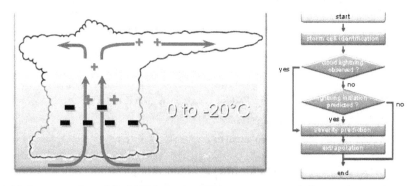

Fig. 10. (a) Conceptual model of CG lightning. The main source of electric charges is assumed to be located in the mixed-phase layer between 0 and -20°C. Prior to electrification, the updraft is expected to separate the charge carriers vertically. Negative charge carriers (i.e. graupel) are expected to reside mainly in the mixed-phase layer. The updraft pumps super-cooled rain water into this layer and wet the carriers. (b) Flow chart of the logic of the algorithm.

SWIRLS updates and outputs nowcast products at 6-minute intervals. For severe thunderstorms, the major results are visualized as an image product called the Severe Weather Map on its client workstation in the forecasting office, as well as a web page named SPIDASS (**S**WIRLS **P**anel for **I**ntegrated **D**isplay of **A**lerts on **S**evere **S**torms) dedicated for severe weather alerts (Fig. 11).

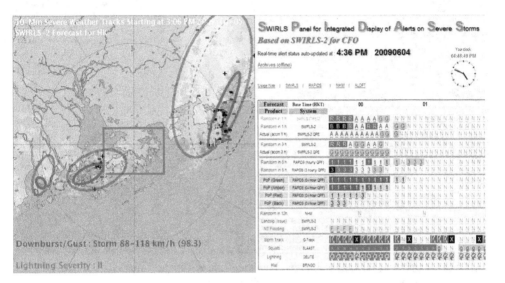

Fig. 11. (a) shows an example of the Severe Weather Map. Textual alerts with quantitative details were printed at the bottom. (b) , the main panel of SPIDASS web page provides a compact view of all alerts arranged in rows and colour-coded for different severity levels.

The Hong Kong Observatory has also developed separate multi-sensor thunderstorm nowcasting systems for the aviation community and the public utilities services (Li, 2009). A lightning nowcasting system, named the Airport Thunderstorm and Lightning Alerting System (ATLAS), covers the Hong Kong International Airport (HKIA). It combines rapidly updated CG lightning strike information, radar reflectivity and TREC wind information to nowcast lightning strikes using a modified Semi-Lagrangian advection scheme. Depending on the predicted distance from HKIA, ATLAS will automatically generate RED (1km) or AMBER (5 km) alerts.

ATLAS is equipped with two ensemble algorithms, to take into account the possible rapid development nature of lightning (transient and sporadic). The Weighted Ensemble (WE) algorithm sums all available 12-minute CG forecasts with decreasing weight with time. If the sum exceeds an optimized threshold, alerts are created. WE has proved to be effective for alerting persistent and wide-spread thunderstorms. The Time Lagged Ensemble (TLE) algorithm sums the 1-minute forecasts valid at the same time from the twelve 1-minute forecasts provided in the past 12 minutes with decreasing weight over time. TLE is proved to be more skilful in predicting rapidly developing, small or wide-spread thunderstorms than WE. Figure 12 shows a snapshot of the ATLAS product.

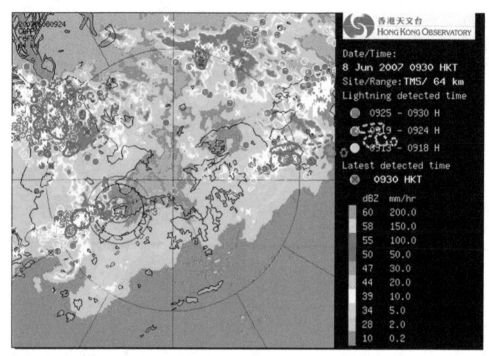

Fig. 12. A snapshot of ATLAS webpage. The image shows the actual position of the CGs (ellipses with solid line), the predicted CGs (ellipses with dashed line), the 12-minute forecast in blue and the 30-minute forecast in grey.

The Aviation Thunderstorm Nowcasting System (ATNS) has been developed to predict the movement of thunderstorms to help local Air Traffic Management to better manage the flight traffic over the Hong Kong Flight Information Region for the next few hours (Li and Wong, 2010). A blending approach is adopted to extend the forecast range and to capture the development and dissipation of thunderstorms. The NWP model used is a high resolution non-hydrostatic model with horizontal resolution of 5 km (Li et al. 2005; Wong et al. 2009). Volume radar reflectivity data are ingested into the model via the LAPS data assimilation system (Albers et al. 1996) and radar Doppler radial wind and 3D radar winds are assimilated via the JNoVA-3DVAR data assimilation system (Honda et al. 2005) to improve the initial moisture field and wind fields, respectively.

The blending algorithm is as follows: (i) SWIRLS radar forecast reflectivity is converted into surface precipitation using a dynamic reflectivity-rainfall (Z-R) relation; (ii) precipitation forecasts are extracted from the NHM; and (iii) then they are blended. The latter blending process involves: (i) Phase correction where a variational technique minimizes the root mean square error of the forecast rainfall field from a previous model run (usually initialized at 1-2 hours before) and the actual radar-raingauge derived precipitation distribution (Wong et al. 2009). (ii) Calibration of the QPF rainfall intensities is based on the observed radar-based quantitative precipitation estimate (QPE), and (iii) blending of calibrated model QPF with the radar nowcast out to 6 hours where the weighting is biased to the nowcasts in the early stages and towards the model at the longer lead times. Figure 13 shows the comparison between the effects of ATNS using SWIRLS simple extrapolation (1-6 hours forecasts) and ATNS using SWIRLS-NHM blended forecasts for the case of 4 Jun 2009 (1-6 hour forecasts). The simple TREC extrapolation (left column) overpredicts the rainfall intensities in this case at long lead times (6 hours).

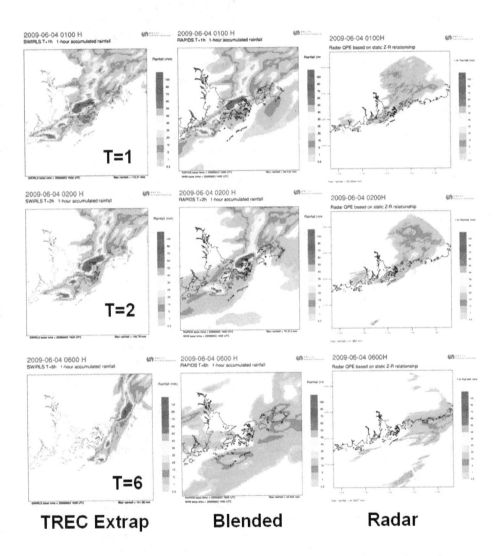

Fig. 13. An example showing the comparison between the effects of SWIRLS simple extrapolation and blending of SWIRLS and NHM rainfall. Figures from top to bottom are 1-hr, 2-hr and 6hr simple extrapolation (left column), AANS blended precipitation (middle column) forecasts and the radar-based QPE (different scale).

6.7 SIGOONS – France

Significant Weather Object Oriented Nowcast System (SIGOONS) is a component of the Synergie workstation (Brovelli et al, 2005). Thunderstorm cells are identified using the RDT (Rapidly Developing Thunderstorms) technique by Hering et al (2005) and are represented as objects. This database is updated every five minutes and is automatically quality controlled against other observational data. The objects may have deterministic and probabilistic attributes and have a time dimension – they can grow and decay. Products are automatically generated and tailored according to pre-defined customer requirements. Discrepancies are brought to the attention of the forecaster who can select persistence over linear extrapolation nowcasts. The forecaster can take additional initiative. The attributes of the weather objects can be manipulated and altered by forecasters.

6.8 THESPA and TIFS, Australia

Within the Bureau of Meteorology, forecasters use RAPIC to interactively interrogate the data. The innovation is the radar data is loaded on the graphics memory of the client computer and extremely rapid response of the display is achieved. To avoid dual-PRF dealiasing errors, only single PRF data is used resulting in a Nyquist interval of 16 m/s. This implies considerable forecaster training is required to interpret highly aliased Doppler data.

Thunderstorm Strike Probability (THESPA, Dance et al, 2010) generates probabilistic nowcasts. Using the historical statistics of the nowcast position errors as a function of lead time and detected storm properties, storm motion is modeled as a bivariate Gaussian distribution on storm speed and direction. For a given geographical point, the strike probability from all possible thunderstorms is computed for the forecast period (Fig. 14).

The algorithm is embedded in the Thunderstorm Interactive Forecast System (TIFS, Bally 2004). The Beijing Olympics provided an opportunity to explore and prototype new nowcasting techniques (Wang et al, 2010). TIFS was modified to ingest the storm locations and tracks from the CARDS, SWIRLS, WDSS and TITAN to create a poor man's ensemble. From each of the storms and tracks, THESPA was used to compute a consensus or ensemble strike probability (Fig. 14b). A warning product would be automatically generated. The analyst (B08 Forecast Demonstration Project team member) would evaluate the product and determine if intervention was needed. The analyst could then use the graphical interface and add, delete or modify cells or tracks. The analyst could view and modify any of the ensemble members and the strike probability display would update. Accepting the change would regenerate the automated warning product, be disseminated and overwriting the fully automated product.

6.9 NoCAWS – SMB

The Shanghai Meteorological Bureau's NoCAWS system was one of the nowcast systems used for the World Expo on Nowcasting Services (WENS) component of the Multi-hazard Early Warning Service project (MHEWS). It integrates observations, mesoscale models and nowcasts to host data displays; analysis tools, severe weather alerting tools to generate automatic forecasts and warning for forecasters. It covers the scales from outlooks to warnings. An innovative feature is lightning forecasts. COTREC winds are used to nowcast cell motions. Advection and statistical relationships between lightning and reflectivity are used to nowcast lightning (Fig. 15)

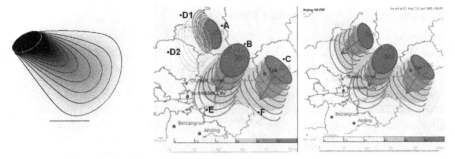

Fig. 14. (a) The detected thunderstorm is the ellipse oriented south-west to north-east. A motion to the south east is shown. The contours and shading show the probability that the thunderstorm will advect or propagate into those locations. The probabilities were verified for a season of storms around Sydney and Beijing, with excellent reliability, with a Brier skill score of between 0.36 and 0.44 with respect to an advected threat area forecast. (b) An example of a prototype TIFS strike probability product. Three cells are identified as A, B and C and represented as ellipses. The tracks of B and C are indicted by the partial ellipses and the colours indicate the strike probability, marked as E and F and appear consistent. The track for cell A is marked as D1 and appears anomalous. D2 is the track that the analyst has modified to produce the final strike probability map (c).

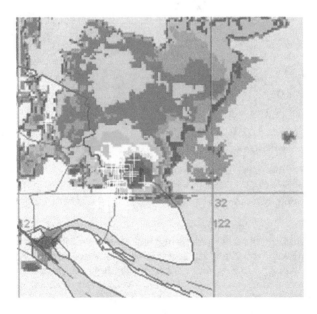

Fig. 15. This figure shows a nowcast of the reflectivity and lightning from NoCAWS. The plus signs are nowcasts of lightning strikes.

6.10 KONRAD/NinJo/NowCastMIX – DWD

There are several tools in the German Weather Service and include KONRAD (Lang et al, 2001), Mesocyclone detection (Hengstebeck et al, 2011), AutoWARN, EPM (editing, prediction, monitoring), Cellviews (Joe et al, 2003). All of these are integrated into the Ninjo system (Koppert et al, 2004). KONRAD was developed as a research prototype and uses a variable elevation angle PPI reflectivity product for the identification and warning potential of cells. The 10 minute volume scan product is used for further classification. The cells are displayed as abstractions and only a >28dBZ contour is displayed in the end user product (Fig. 16). Of all the systems discussed, it is the only truly automated system where the products go directly out to the end-user without human oversight. However, it targets sophisticated end-users such as emergency authorities, county administrators, fire departments and the military and not the public. One could argue that these are guidance products for external versus internal decision-makers for planning but not warning service. So the "cry wolf" syndrome is not a significant issue. This does demonstrate the potential use of fully authomated products.

From echoes to symbols

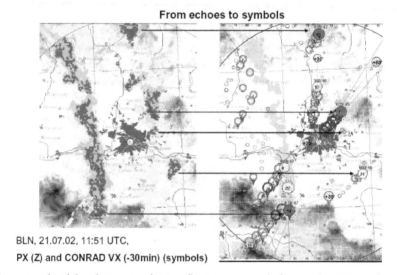

BLN, 21.07.02, 11:51 UTC,
PX (Z) and CONRAD VX (-30min) (symbols)

Fig. 16. An example of the abstraction from reflectivity to symbolic representation of thunderstorms from the KONRAD system. It is the only system described in the contribution that is totally automated. It is directed to "sophisticated users" for planning purposes.

The AutoWARN system in NinJo integrates various meteorological data and products in a warning decision support process, generating real-time warning proposals for assessment and possible modification by the duty forecasters. These warnings finally issued by the forecaster are then exported to a system generating textual and graphical warning products for dissemination to customers. On very short, nowcasting timescales, several systems are continuously monitored. These include the radar-based storm-cell identification and tracking methods, KONRAD and CellMOS; 3D radar volume scans yielding vertically integrated liquid water (VIL) composites; precise lightning strike locations; the precipitation prediction system, RadVOR-OP as well as synoptic reports and the latest high resolution numerical analysis and forecast data.

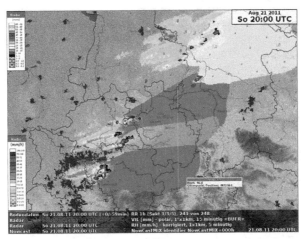

Fig. 17. An example of NowcastMix. It combines and merges the output from several nowcasting systems into a hazard map.

Since there are several nowcasting systems avaliable, NowCastMIX processes these available nowcast products together in an integrated grid-based analysis, providing a generic, optimal warning solution with a 5-minute update cycle. The products are combined using a fuzzy logic approach (James et al 2011). The method includes estimates for the storm cell motion by combining raw cell tracking inputs from the KONRAD and CellMOS systems with vector fields derived from comparing consecutive radar images. Finally, the resulting gridded warning fields are spatially filtered to provide regionally-optimized warning levels for differing thunderstorm severities for forecasters. NowCastMIX delivers a synthesis of the various nowcasting and forecast model system inputs to provide consolidated sets of most-probable short-term forecasts (Fig. 17).

6.11 Japan – JMA

Japan Meteorological Agency initiated their hazardous wind warning program in 2007. A hazardous-wind-possibility-index is calculated based on the NWP prediction of wind and radar reflectivity exceeding a threshold. An innovation is the use of a template matching technique for the detection of mesocyclones. Rankine vortex and divergence flow field templates of different intensity and spatial scale are generated and matched to the radial velocity field. This is done every five minutes. Detections on two consecutive time steps are required as a quality controlled metric. Then the two estimates are combined every ten minutes to estimate a hazardous wind potential. Nowcasting is based on a motion analysis. Different thresholds are statistically established and the success ratio (1-FAR) and the probability of detection (POD) are used to categorize the hazard level (Table 7). If level 2 is exceeded (see Fig. 18), then it alerts a forecaster to issue Hazardous Wind Watch. A forecaster may ignore the level 2 information, when: (i) the storm is near the boundary of a warning area and it will be out before the time of warning or (ii) the quality of radar data seems poor (e.g. AP or sea clutter). A forecaster can issue a warning at level 1 when (i) reliable report of a tornado/tornadoes and/or and (ii) strong gust (say, greater than 30 m/s) caused by a convective cloud.

Warning Level	Criteria
2	Success Ratio = 1- FAR = 5- 10% with POD=20-30%
1	1-FAR= 1-5% and POD=60-70%

Table 7. Hazardous Criteria Level

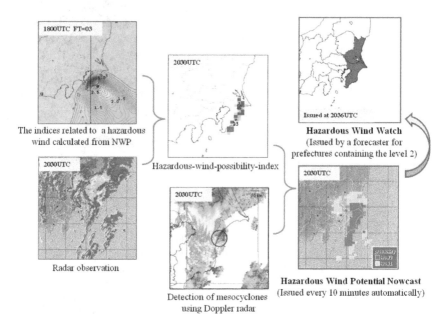

Fig. 18. The processing steps for hazardous wind potential at JMA. It is typical of current systems where mesoscale NWP predictions are assumed to be good enough to match with the observations.

6.12 SWAN – CHINA/CMA

In 2008, the China Meteorological Administration (CMA) launched a campaign on the development of its first version of integrated nowcasting system SWAN (Severe Weather Analysis and Nowcast system). This system aims at providing an integrated, state-of-the-art and timely severe weather nowcast platform for operational forecasters at all levels over China. SWAN ingests data from China's new generation Doppler radars (both S-band and C-band), automatic weather station, satellite, and mesoscale numerical weather prediction model. It offers a tool for severe weather monitoring, analysis, nowcasting and warnings such as flashing a real-time alert, driving next algorithm processes and sending a warning via SMS, etc.

The server application includes several modules, such as providing log files for monitoring system behavior, configuring network environment, setting data acquisition parameters, performing quality control for radar data and AWS data, generating 3D radar reflectivity mosaic, running algorithm for nowcast products, analyzing observation data and providing message for alerting the forecasters.

The client refreshes real-time observations cycled in 5 min from radars and AWS (automatic weather stations) and provides real-time alerts (sounding, flashing) for indication of severe weather events (meeting certain thresholds such as wind speeds or rainfall amounts). It also provides an interactive tool for preparing, editing and issuing Nowcast and warning for severe weathers

Based on quality control, a regional 3D reflectivity mosaic is produced by trying to fill the gaps that are generated by terrain blockage or AP. Products such as vertically integrated liquid (VIL) , echo top (ET) and COTREC winds are then derived. QPE algorithm involves extraction of convective echoes from stratiform echoes by texture and horizontal gradient properties. Different Z-R relations are used for convective rain and stratiform rain. COTREC (continuous tracking radar echo by correlation) vectors are echo motion vectors that are derived from moving radar reflectivity patterns through grid-to-grid cross-correlation and then adjusted by a horizontal non-divergence constraint for hourly nowcasts of rainfall (Li et al, 1995). This is blended with mesoscale numerical prediction model output for 2-3 hour nowcasts.

SWAN provides real time verifications for storm tracking and reflectivity nowcasts. Storm track errors are shown as distance differences between observed storm tracks and predicted storm tracks (1h). Observed radar reflectivity are also verified against extrapolated forecasted reflectivity.

Severe weather warnings can be prepared and issued through SWAN by graphical interface by circling an area on the screen, clicking an icon and doing some minor wording (Fig. 19). A web-based version of SWAN has been developed and deployed in Guangdong Meteorological Bureau.

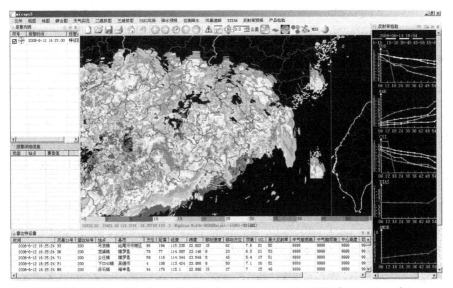

Fig. 19. A SWAN display showing cells/tracks (main screen), SCIT (bottom) and time histories of critical parameters (right). There are similarities with WDSS, NinJo and CARDS displays.

7. Conclusion

The objective of this contribution was to provide a broad overview of the use of radar and radar networks for the provision of severe weather warnings and to very briefly describe historical legacies and current practice. The target audience are those NHMS' who might be contemplating developing or enhancing such a service. Weather radar clearly plays a central role in this application. Not discussed are important applications such as nowcasting precipitation, quantitiative precipitation estimation, wind retrieval, data assimilation for numerical weather prediction, etc. It also does not address the convective initiation aspects (Roberts et al, 2006; Sun et al, 1991; Sun and Crook, 1994). For a reliable warning service, design, infrastructure (reliable power and telecommunications), support and maintenance are critical and were not discussed in this contribution. These are major considerations but out of scope for this contribution.

The level and nature of the service will be determined by both meteorological and non-meteorological factors. The prevalence of severe weather, climatology and a defining event determine the impact, the exposure and the opportunity to develop a warning service. Socio-economic factors, risk persona, as well as the organizational structure, are particularly important in the design and expectations for the radar processing, visualization and dissemination systems. This contribution provided a short global survey of radar based systems to illustrate the commonality but also the differences in implementation. One solution does not fit all. Underlying these systems is the forecast process and it is emphasized that they all rely on human expertise in the decision-making process and so the human-machince mix is a critical item. This will drive the expertise and therefore the training requirements for the severe weather analyst.

This contribution highlighted the use of automation in the production of guidance products. Some systems rely on very little automation and totally rely on manual interpretation. All systems, except one, default to this mode. One of most highly automated systems is CARDS (Canada). Automation is necessary because of the need for look at details for warning preparation purposes while maintaining situational awareness in the situation where one forecaster is responsible for about ten radars. It processes radar data for identifying and ranking thunderstorm cells and features. It also creates highly processed image products to streamline and to guide the decision-making process. It still relies on human decision-making for the final preparation of the warning. KONRAD is the only system that produces totally automated products. However, it could be argued that these products are directed to "sophisticated users" for their specific planning and decision-making purposes and not warning purposes.

Given the limited space and time, all radar processing systems were inadequately described. There is room for improvement in describing all aspects of the processing chain from better algorithms (e.g. hail, hook echoes; Lemon, 1998; Wang et al, 2011) to advanced concepts where thermodynamic diagnostic fields, useful for understanding, are retrieved (Sun et al, 1991; Sun and Crook, 1994). Through the description of specific innovative aspects of individual systems, and since there are commonalities amongst them, the intent was to provide the reader with an overview of the capabilities of all the systems. There is fine work being done elsewhere that is not represented; to name a few, Italy, Switzerland and Finland. Another glaring oversight is the lack of description of systems by manufacturers. Some even offer the possibility for the NHMS to add their own specialized products into their systems. Many of the countries mentioned above in fact use a combination of products from their own systems and those of the manufacturers. Information is readily available in trade shows or on their web sites. The ideal requirement is a seamless, user-friendly integrated visualization, decision-making and production system to cover all scales (the seamless prediction concept) and this is the trend in many NHMS' for all data, products and so radar only processing or visualization systems are an interim step towards this and requires investment, resources, time and effort to achieve. NinJo and AWIPS (not described here) provides an example of how radar is expected to be integrated into a comprehensive forecast analysis, diagnosis, prognosis and production tool.

The purpose of this contribution was to illustrate the issues faced by NHMS's. There is a push to use meteorological technology as much as possible and to automate as much as possible. Computing technology is still a limiting factor – computers, telecommunications and data/product storage are all continuing issues that can always be faster and bigger. If there is the time, the resources and the expertise, manual interpretation of basic radar products is still the best way to provide severe weather warning services and to optimally utilize the considerable capabilities of the forecasters. However, tools are needed to streamline and accelerate the process but this is highly dependent on organizational factors. Automated products introduce another level of complexity and knowledge requirement. They can be black boxes that bewilder the user. However, creating black boxes without diagnostic capabilities, providing poor tools and denying access to basic products and information, is self-defeating. It is a sure way of making smart people (appear) "dumb". The algorithms aren't perfect given the need for high POD. They never will be and they can be better and substantial work on data quality, feature detection and prediction are needed. The systems described exhibit the great efforts and resources are expended to do this. Saving a single button click or a mouse movement can make the difference between a bad and a good system. This is difficult to describe as a requirement and prototyping and demonstration projects are the only way to appreciate this.

While reliable weather radars and expertise play a central role in the warning process, this is still a challenge for many countries. Satellite and lightning systems are now available that have minimal support requirements. Stand alone applications for severe weather can and are being developed for these system. In the absence of radars, there is no question that they will provide benefits but their efficacy, the forecast process and the service level for severe weather warnings need to be demonstrated. No doubt that they should also enhance existing systems that rely on weather radar networks. This is occurring but beyond the scope of this contribution. No convective scale warning service has been soley developed without radar and so this is a new area to investigate. Understanding the technology, interpretation of the data and the products will require more development, enhanced expertise, demonstration and decision-making skills.

For the convective weather problem, dual-polarization radar will have benefits in data quality, hail detection and rainfall estimation but this is again beyond the scope of this contribution (Frame et al, 2009). Earth curvature and beam propagation preclude low level detection and so many of the hazardous phenomena are not actually measured beyond a few tens of kilometer from the radar site and must therefore be inferred from measurements aloft. The CASA (Cooperative Adapting and Sensing of the Atmosphere) is a network of X

Band radars that address this issue but it is in early-transitional development (McLaughlin et al, 2009; Ruzanski et al, 2011). It also addresses the issue of rapid or adapting scan strategies (Heinselman et al, 2008) which is being investigated now but beyond the scope of this contribution. In any case, with increasing computing power, telecommunications, additional observations and new technology, these are exciting times.

8. References

AMS, 2001: Expectations Concerning Media Performance during Severe Weather Emergencies (Adopted by the AMS Council 14 January 2001), Bulletin of the American Meteorological Society, 2001: Volume 82, Issue 4 705-70.

Albers S., J. McGinley, D. Birkenheuer, and J. Smart 1996: The Local Analysis and Prediction System (LAPS): Analyses of clouds, precipitation, and temperature, Wea. Forecasting, 11, 273-287.

Andra Jr., D. L., E. M. Quoetone, W. F. Bunting, 2002: Warning Decision Making: The Relative Roles of Conceptual Models, Technology, Strategy, and Forecaster Expertise on 3 May 1999, Weather and Forecasting, Volume 17, Issue 3 (June 2002) 559-566.

Bally, J., 2004: The Thunderstorm Interactive Forecast System: Turning Automated Thunderstorm Tracks into Severe Weather Warnings, Weather and Forecasting, Volume 19, Issue 1 (February 2004) 64-7.

Barnes, L. R., E. C. Gruntfest, M. H. Hayden, D. M. Schultz, C. Benight, 2007: False Alarms and Close Calls: A Conceptual Model of Warning Accuracy, Weather and Forecasting, Volume 22, Issue 5 (October 2007) 1140-114.

Baumgart, L. A., E. J. Bass, B. Philips, K. Kloesel, 2008: Emergency Management Decision Making during Severe Weather, Weather and Forecasting, Volume 23, Issue 6 (December 2008) 1268-127.

Bech, J., Vilaclara E., Pineda, N., Rigo, T., Lopez, J., O'Hora, F., Lorente, J., Sempere, D., Fabregas F.X., 2004: The weather radar network of the Catalan meteorological service: description and applications, ERAD, Visby, Sweden, 416-420.

Bellon, A. and G. L. Austin, 1978: The Evaluation of Two Years of Real-Time Operation of a Short-Term Precipitation Forecasting Procedure (SHARP), Journal of Applied Meteorology, Volume 17, Issue 12, 1778-1787.

Bieringer, P., P. S. Ray, 1996: Comparison of Tornado Warning Lead Times with and without NEXRAD Doppler Radar, Weather and Forecasting, Volume 11, Issue 1 (March 1996) 47-5.

Black, A. W., W. S. Ashley, 2011: The Relationship between Tornadic and Nontornadic Convective Wind Fatalities and Warnings, Weather, Climate, and Society, Volume 3, Issue 1 (January 2011) 31-4.

Branick, M. L., C. A. Doswell III, 1992: Polarity, An Observation of the Relationship between Supercell Structure and Lightning Ground-Strike, Weather and Forecasting, Volume 7, Issue 1) 143-14.

Breidenbach, J. P., D. H. Kitzmiller, W. E. McGovern, R. E. Saffle, 1995: The Use of Volumetric Radar Reflectivity Predictors in the Development of a Second-Generation Severe Weather Potential Algorithm, Weather and Forecasting, Volume 10, Issue 2 (June 1995) 369-37.

Brooks, H. E., C. A. Doswell III, J. Cooper, 1994: On the Environments of Tornadic and Nontornadic Mesocyclones, Weather and Forecasting, Volume 9, Issue 4 (December 1994) 606-61.

Brooks, H. E., C. A. Doswell III, L. J. Wicker, 1993: STORMTIPE: A Forecasting Experiment Using a Three-Dimensional Cloud Model, Weather and Forecasting, Volume 8, Issue 3 (September 1993) 352-36.

Brooks, H. E., C. A. Doswell III, R. B. Wilhelmson, 1994: The Role of Midtropospheric Winds in the Evolution and Maintenance of Low-Level Mesocyclones, Monthly Weather Review, Volume 122, Issue 1 (January 1994) 126-13.

Brovelli, P., S. Sénési, E. Arbogast, P. Cau, S. Cazabat, M. Bouzom, J. Reynaud, 2005: Nowcasting thunderstoms with SIGOONS. A significant weather object oriented nowcasting system, Météo-France, Toulouse, France, WSN05

Brown, R. A., J. M. Janish, V. T. Wood, 2000: Impact of WSR-88D Scanning Strategies on Severe Storm Algorithms, Weather and Forecasting, Volume 15, Issue 1 (February 2000) 90-10.

Brunner, J. C., S. A. Ackerman, A. S. Bachmeier, R. M. Rabin, 2007: A Quantitative Analysis of the Enhanced-V Feature in Relation to Severe Weather, Weather and Forecasting, Volume 22, Issue 4 (August 2007) 853-87.

Burgess, D. W., R. J. Donaldson JR., and P. R. Desrochers, 1993: Tornado detection and warning by radaR. The Tornado: Its Structure, Dynamics, Prediction, and Hazards, Geophys. Monogr., No. 79, Amer. Geophys. Union, 203–221.

Byko, Z., P. Markowski, Y. Richardson, J. Wurman, E. Adlerman, 2009: Descending Reflectivity Cores in Supercell Thunderstorms Observed by Mobile Radars and in a High-Resolution Numerical Simulation, Weather and Forecasting, Volume 24, Issue 1 (February 2009) 155-18.

Chisholm, A. J. and J. Renick, 1972: The kinematics of multicell and supercell Alberta hailstorms, Alberta Hail Studies, 1972, Reseach Council of Alberta Hail Studies Rep. No. 72-2, 24-31.

Crane, R. K., 1979: Automatic cell detection and tracking. IEEE., Trans. Geosci. Electron., GE-17, 250–262.

Crum, T. D. and R. L. Alberty, 1993: The WSR-88D and the WSR-88D Operational Support Facility, Bulletin of the American Meteorological Society, Volume 74, Issue 9, 1669-1687.

Dance, S., E. Ebert, D. Scurrah, 2010: Thunderstorm Strike Probability Nowcasting, Journal of Atmospheric and Oceanic Technology, Volume 27, Issue 1) 79-9.

Dance, S., R. Potts, 2002: Microburst Detection Using Agent Networks, Journal of Atmospheric and Oceanic Technology, Volume 19, Issue 5) 646-65.

Davis, C., N. Atkins, D. Bartels, L. Bosart, M. Coniglio, G. Bryan, W. Cotton, D. Dowell, B. Jewett, R. Johns, D. Jorgensen, J. Knievel, K. Knupp, W. C. Lee, G. Mcfarquhar, J. Moore, R. Przybylinski, R. Rauber, B. Smull, R. Trapp, S. Trier, R. Wakimoto, M. Weisman, C. Ziegler, 2004: The Bow Echo and MCV Experiment: Observations and Opportunities, Bulletin of the American Meteorological Society, Volume 85, Issue 8 (August 2004) 1075-1093

Dixon, M. and G. Weiner, 1993: TITAN, Thunderstorm Identification, Tracking, Analysis and Nowcasting - A Radar-based Methodology, JAOT, 10, 785-797.

Donaldson, Jr., Ralph J., P. R. Desrochers, 1990: Improvement of Tornado Warnings by Doppler Radar Measurement of Mesocyclone Rotational Kinetic Energy, Weather and Forecasting, Volume 5, Issue 2 (June 1990) 247-25.

Doswell III, C. A., 1980: Synoptic-Scale Environments Associated with High Plains Severe Thunderstorms, Bulletin of the American Meteorological Society, Volume 61, Issue 11 (November 1980) 1388-1400.

Doswell, C. A. III, 1982: The operational meteorology of convective weather, Volume I: Operational Mesoanalysis, NOAA Tech Memo ERL NSSFC-5, 168p.

Doswell, C. A. III, 1985: The operational meteorology of convective weather, Volume II: Storm Scale Analysis, NOAA Tech Memo ERL ESG-15, 240p.

Doswell III, C. A., 2001: Severe Convective Storms, Meteorological Monographs, Volume 28, 561pp.

Doswell III, C. A. 2004: Weather Forecasting by Humans—Heuristics and Decision Making. Weather and Forecasting 19:6, 1115-112.

Doswell III, C. A., R. Davies-Jones, D. L. Keller, 1990: On Summary Measures of Skill in Rare Event Forecasting Based on Contingency Tables, Weather and Forecasting, Volume 5, Issue 4 (December 1990) 576-58.

Doswell III, C. A., R. Edwards, R. L. Thompson, J. A. Hart, K. C. Crosbie, 2006: A Simple and Flexible Method for Ranking Severe Weather Events, Weather and Forecasting, Volume 21, Issue 6 (December 2006) 939-95.

Doswell III, C. A., A. R. Moller, H. E. Brooks, 1999: Storm Spotting and Public Awareness since the First Tornado Forecasts of 1948, Weather and Forecasting, Volume 14, Issue 4 (August 1999) 544-557.

Doviak, R. J., and D. S. Zrnic, 1984: Doppler Radar and weather observations, Academic Press, 458p.

Dunn, L. B., 1990: Two Examples of Operational Tornado Warnings Using Doppler Radar Data, Bulletin of the American Meteorological Society, Volume 71, Issue 2 (February 1990) 145-15.

Ebert, E. E., L. J. Wilson, B. G. Brown, P. Nurmi, H. E. Brooks, J. Bally, M. Jaeneke, 2004: Verification of Nowcasts from the WWRP Sydney 2000 Forecast Demonstration Project, Weather and Forecasting, Volume 19, Issue 1 (February 2004) 73-96

Eilts, M. D., and Coauthors, 1996: Severe weather warning decision support system. Preprints, 18th Conf. on Severe Local Storms, San Francisco, CA, AmeR. MeteoR. Soc., 536-540.

Evans, J. S., C. A. Doswell III, 2001: Examination of Derecho Environments Using Proximity Soundings, Weather and Forecasting, Volume 16, Issue 3 (June 2001) 329-34.

Fawbush, E. J. and R. C. Miller, 1953: Forecasting Tornadoes, USAF Air University Quarterly Review, 1, 108-11.

Fox, Neil I., R. Webb, J. Bally, M. W. Sleigh, C. E. Pierce, D. M. L. Sills, P. I. Joe, J. Wilson, C. G. Collier, 2004: The Impact of Advanced Nowcasting Systems on Severe Weather Warning during the Sydney 2000 Forecast Demonstration Project: 3 November 2000, Weather and Forecasting, Volume 19, Issue 1 (February 2004) 97-114

Frame, J., P. Markowski, Y. Richardson, J. Straka, J. Wurman, 2009: Polarimetric and Dual-Doppler Radar Observations of the Lipscomb County, Texas, Supercell Thunderstorm on 23 May 2002, Monthly Weather Review, Volume 137, Issue 2 (February 2009) 544-56.

Galway, J. G., 1989: The Evolution of Severe Thunderstorm Criteria within the Weather Service, Weather and Forecasting, Volume 4, Issue 4 (December 1989) 585-59.

Gatlin, P. N., S. J. Goodman, 2010: A Total Lightning Trending Algorithm to Identify Severe Thunderstorms, Journal of Atmospheric and Oceanic Technology, Volume 27, Issue 1 (January 2010) 3-22.

Glahn, B., 2005: Tornado-Warning Performance in the Past and Future—Another Perspective, Bulletin of the American Meteorological Society, Volume 86, Issue 8 (August 2005) 1135-114.

Goodman, S. J., D. E. Buechler, P. J. Meyer, 1988: Convective Tendency Images Derived from a Combination of Lightning and Satellite Data, Weather and Forecasting, Volume 3, Issue 3 (September 1988) 173-188.

Hammer, B., T. W. Schmidlin, 2002: Response to Warnings during the 3 May 1999 Oklahoma City Tornado: Reasons and Relative Injury Rates, Weather and Forecasting, Volume 17, Issue 3 (June 2002) 577-58.

Heinselman, P. L., D. L. Priegnitz, K. L. Manross, T. M. Smith, R. W. Adams, 2008: Rapid Sampling of Severe Storms by the National Weather Radar Testbed Phased Array Radar, Weather and Forecasting, Volume 23, Issue 5 (October 2008) 808-824.

Hengstebeck, T., D. Heizenreder, P. Joe, P. Lang, 2011: The Mesocyclone Detection Algorithm of DWD, 6th European Conference on Severe Storms, ECSS, 3-7 October 2011, Palma de Mallorca

Hermes, L. G., A. Witt, S. D. Smith, D. Klingle-Wilson, D. Morris, G. J. Stumpf, M. D. Eilts, 1993: The Gust-Front Detection and Wind-Shift Algorithms for the Terminal Doppler Weather Radar System, Journal of Atmospheric and Oceanic Technology, Volume 10, Issue 5 (October 1993) 693-70.

Hering, A. M., S. Senesi, P. Ambrosetti and I. Bernard-Bouissieres, 2005: Nowcasting thunderstorms in complex cases using radar data, WMO Symposium on Nowcasting and Very Short Range Forecasting, Toulouse France, paper 2. 14.

Hoekstra, S., K. Klockow, R. Riley, J. Brotzge, H. Brooks, S. Erickson, 2011: A Preliminary Look at the Social Perspective of Warn-on-Forecast: Preferred Tornado Warning Lead Time and the General Public's Perceptions of Weather Risks, Weather, Climate, and Society, Volume 3, Issue 2 (April 2011) 128-14.

Holleman, I. and H. Beekhuis, 2003: Analysis and correction of dual PRF velocity data, JAOT, 20(4), 443-453.

Honda, Y., M. Nishijima, K. Koizumi, Y. Ohta, K. Tamiya, T. Kawabata and T. Tsuyuki, 2005: A pre-operational variational data assimilation system for a non-hydrostatic model at the Japan Meteorological Agency: Formulation and preliminary results. Quart. J. Roy. Meteor. Soc., 131, 3465-3475.

James, P. M., S. Treple, D. Heizenreder and B. K. Reichert, 2011: NowCastMIX – A fuzzy logic based tool for providing automatic integrated nowcasting systems, 11th EMS Annual Meeting, 10th European Conference on Applications of Meteorology, 12-16 Sept 2011., EMS2011-234,

Joe, P., 2009: A First Look at Radar Data Quality for the Beijing 2008 Forecast Demonstration Project, in Collection of Papers on the New Generation of China Radars, (Xin Yi Dai, Tian Qi Leida Yewu Yingyoung Lunwenji, ISBN 978-7-5029-4468-1), edited by Xaioding Yu (invited lead paper).

Joe, P., D. Burgess, R. Potts, T. Keenan, G. Stumpf, A. Treloar, 2004: The S2K Severe Weather Detection Algorithms and Their Performance, Weather and Forecasting, Volume 19, Issue 1, 43-63.

Joe, P., M. Falla, P. Van Rijn, L. Stamadianos, T. Falla, D. Magosse, L. Ing and J. Dobson, 2002: Radar Data Processing for Severe Weather in the National Radar Project of Canada, SELS, San Antonio, 12-16 August 2002, 221-224.

Joe, P., P. T. May, 2003: Correction of Dual PRF Velocity Errors for Operational Doppler Weather Radars, Journal of Atmospheric and Oceanic Technology, Volume 20, Issue 4 (April 2003) 429-44.

Johns, R. H., C. A. Doswell III, 1992: Severe Local Storms Forecasting, Weather and Forecasting, Volume 7, Issue 4 (December 1992) 588-61.

Johnson, J. T., P. L. MacKeen, A. Witt, E. De W. Mitchell, G. J. Stumpf, M. D. Eilts, K. W. Thomas., 1998: The Storm Cell Identification and Tracking Algorithm: An Enhanced WSR-88D Algorithm, Weather and Forecasting, Volume 13, Issue 2, 263-27.

Keenan, T., P. Joe, J. Wilson, C. Collier, B. Golding, D. Burgess, P. May, C. Pierce, J. Bally, A. Crook, A. Seed, D. Sills, L. Berry, R. Potts, I. Bell, N. Fox, E. Ebert, M. Eilts, K. O'Loughlin, R. Webb, R. Carbone, K. Browning, R. Roberts, C. Mueller, 2004: The Sydney 2000 World Weather Research Programme Forecast Demonstration Project: Overview and Current Status, Bulletin of the American Meteorological Society, Volume 84, Issue 8, 1041-1054.

Kessler, E., J. W. Wilson, 1971: Radar in an Automated National Weather System, Bulletin of the American Meteorological Society, Volume 52, Issue 11, 1062-106.

King, P. W. S., M. J. Leduc, D. M. L. Sills, N. R. Donaldson, D. R. Hudak, P. Joe, B. P. Murphy, 2003: Lake Breezes in Southern Ontario and Their Relation to Tornado Climatology, Weather and Forecasting, Volume 18, Issue 5 (October 2003) 795-807.

Kitzmiller, D. H., W. E. McGovern, R. F. Saffle, 1995: The WSR-88D Severe Weather Potential Algorithm, Weather and Forecasting, Volume 10, Issue 1 (March 1995) 141-15.

Klingle, D. L., D. R. Smith, M. M. Wolfson, 1987: Gust Front Characteristics as Detected by Doppler Radar, Monthly Weather Review, Volume 115, Issue 5 (May 1987) 905-91.

Knupp, K. R., S. Paech, S. Goodman, 2003: Variations in Cloud-to-Ground Lightning Characteristics among Three Adjacent Tornadic Supercell Storms over the Tennessee Valley Region, Monthly Weather Review, Volume 131, Issue 1 (January 2003) 172-188

Koppert, H. -J., Pedersen, T. S., Zuercher, B., Joe, P., 2004: How to make an international Meteorological Workstation project successful, BAMS, 1087-109.

Lakshmanan, J. T. ., A. Fritz, T. Smith, K. Hondl, and G. J. Stumpf, 2007: An automated technique to quality control radar reflectivity data, J. Applied Meteorology, vol. 46, 288-305.

Lakshmanan, V., K. Hondl, and R. Rabin, 2009: An efficient, general-purpose technique for identifying storm cells in geospatial images, J. Ocean. Atmos. Tech., vol. 26, no. 3, 523-37.

Lakshmanan, V., R. Rabin, and V. DeBrunner, 2003: Multiscale storm identification and forecast, J. Atm. Res., vol. 67, 367-380.

Lakshmanan, V. and T. Smith, 2009: Data mining storm attributes from spatial grids, J. Ocea. and Atmos. Tech., vol. 26, no. 11, 2353-2365.

Lakshmanan, V. and T. Smith, 2010: An Objective Method of Evaluating and Devising Storm-Tracking Algorithms, Weather and Forecasting, Volume 25, 701-709.

Lakshmanan, V., T. Smith, K. Hondl, G. J. Stumpf, A. Witt, 2006: A Real-Time, Three-Dimensional, Rapidly Updating, Heterogeneous Radar Merger Technique for Reflectivity, Velocity, and Derived Products, Weather and Forecasting, Volume 21, Issue 5, 802-82.

Lakshmanan, V., T. Smith, G. Stumpf, K. Hondl, 2007: The Warning Decision Support System–Integrated Information, Weather and Forecasting, Volume 22, Issue 3, 596-61.

Lakshmanan, V., J. Zhang, K. Hondl, and C. Langston, 2011: A statistical approach to mitigating persistent clutter in radar reflectivity data, IEEE J. Selected Topics in Applied Earth Observations and Remote Sensing, vol. s, p. accepted.

Lakshmanan, V., J. Zhang, and K. Howard, 2010, A technique to censor biological echoes in radar reflectivity data, J. Applied Meteorology, vol. 49, 435-462.

Lang, P., 2001: Cell tracking and warning indicators derived from operational radar products, 30th International Radar Conference, AMS, Munich, Germany, 245-247.

Lang, T. J., L. J. Miller, M. Weisman, S. A. Rutledge, L. J. Barker III, V. N. Bringi, V. Chandrasekar, A. Detwiler, N. Doesken, J. Helsdon, C. Knight, P. Krehbiel, Walter A. Lyons, D. Macgorman, E. Rasmussen, W. Rison, W. D. Rust, Ronald J. T., 2004: The Severe Thunderstorm Electrification and Precipitation Study, Bulletin of the American Meteorological Society, Volume 85, Issue 8 (August 2004) 1107-1125

Lapczak, S., E. Aldcroft, M. Stanley-Jones, J. Scott, P. Joe, P. Van Rijn, M. Falla, A. Gagne, P. Ford, K. Reynolds and D. Hudak, 1999: The Canadian National Radar Project, 29th Conf. Radar Met., Montreal, AMS, 327-330.

Leduc, M., P. Joe, M. Falla, P. Van Rijn, S. Lapczak, I. Ruddick, A. Ashton and R. Alsen, 2002: The July 4 2001 Severe Weather Outbreak in Southern Ontario as Diagnosed by the New Radar Data Processing System of the National Radar Project of Canada, SELS, San Antonio, 12-16 August 2002, 170-173.

Lei H., S. Fu, L. Zhao, Y. Zheng, H. Wang and Y. Lin, 2009: 3D Convective Storm Identification, Tracking, and Forecasting — An Enhanced TITAN Algorithm, Journal of Atmospheric and Oceanic Technology, 2009: Volume 26, Issue 4) 719-73.

Lemon, L. R., 1977: new severe thunderstorm radar identification techniques are warning criteria: a preliminary report, NWS NSSFC-1, PB 273049 60p.

Lemon, L. R., 1980: new severe thunderstorm radar identification techniques are warning criteria, NWS NSSFC-1, PB 231409 60p.

Lemon, L. R., 1998: The Radar "Three-Body Scatter Spike": An Operational Large-Hail Signature, Weather and Forecasting, Volume 13, Issue 2 (June 1998) 327-34.

Lemon, L. R., Ralph J. Donaldson, Jr., D. W. Burgess, R. A. Brown, 1977: Doppler Radar Application to Severe Thunderstorm Study and Potential Real-Time Warning, Bulletin of the American Meteorological Society, Volume 58, Issue 11, 1187-119.

Lemon, L. R., C. A. Doswell III, 1979: Severe Thunderstorm Evolution and Mesocyclone Structure as Related to Tornadogenesis, Monthly Weather Review, Volume 107, Issue 9 (September 1979) 1184-119.

Lenning, E., H. E. Fuelberg, A. I. Watson, 1998: An Evaluation of WSR-88D Severe Hail Algorithms along the Northeastern Gulf Coast, Weather and Forecasting, Volume 13, Issue 4 (December 1998) 1029-104.

Li, P.W. 2009: Development of a thunderstorm nowcasting system for Hong Kong International Airport, AMS Aviation, Range, Aerospace Meteorology Special Symposium on Weather-Air Traffic Management Integration, Phoenix, Arizona, 11-15 Jan 2009.

Li, P. W. and W.K. Wong, 2010: Development of an Advanced Aviation Nowcasting System by Including Rapidly Updated NWP Model in Support of Air Traffic Management, Proceedings 14th Conference on Aviation, Range and Aerospace Meteorology, Atlanta, Georgia, USA, 17-21 January 2010.

Li, L., W. Schmid and J. Joss, 1995: Nowcasting of motion and growth of precipitation with radar over a complex orography, JAM, 34(6), 1286-1300

Li, P. W. and S. T. Lai, 2004: Applications of radar-based nowcasting techniques for mesoscale weather forecasting in Hong Kong, Meteorological Applications, 11, 253-264.

Li, P. W. and S. T. Lai, 2004a: Short-range quantitative precipitation forecasting in Hong Kong, Journal of Hydrology, 288, 189-209.

Li, P.W., W.K. Wong and E.S.T. Lai, 2005: A New Thunderstorm Nowcasting System in Hong Kong, WMO/WWRP International Symposium on Nowcasting and Very-short-range Forecasting, Toulouse, France, 5-9 Sep. 2005.

Markowski, P. M., 2002: Hook Echoes and Rear-Flank Downdrafts: A Review, Monthly Weather Review, Volume 130, Issue 4 (April 2002) 852-87.

Markowski, P. M., E. N. Rasmussen, J. M. Straka, 1998a: The Occurrence of Tornadoes in Supercells Interacting with Boundaries during VORTEX-95, Weather and Forecasting, Volume 13, Issue 3 (September 1998) 852-85.

Markowski, P. M., J. M. Straka, E. N. Rasmussen, D. O. Blanchard, 1998b: Variability of Storm-Relative Helicity during VORTEX, Monthly Weather Review, Volume 126, Issue 11 (November 1998) 2959-297.

Marshall, J. S. and E. H. Ballantyne, 1975: Weather Surveillance Radar, J. A. M., 14, 1317-1338.

May, P. T., T. D. Keenan, R. Potts, J. W. Wilson, R. Webb, A. Treloar, E. Spark, S. Lawrence, E. Ebert, J. Bally, P. Joe, 2004: The Sydney 2000 Olympic Games Forecast Demonstration Project: Forecasting, Observing Network Infrastructure, and Data Processing Issues, Weather and Forecasting, Volume 19, Issue 1, 115-130.

McCarthy, J., J. W. Wilson, T. T. Fujita, 1982: The Joint Airport Weather Studies Project, Bulletin of the American Meteorological Society, Volume 63, Issue 1) 15-1.

McLaughlin, D., D. Pepyne, B. Philips, J. Kurose, M. Zink, D. Westbrook, E. Lyons, E. Knapp, A. Hopf, A. Defonzo, R. Contreras, T. Djaferis, E. Insanic, S. Frasier, V. Chandrasekar, F. Junyent, N. Bharadwaj, Y. Wang, Y. Liu, B. Dolan, K. Droegemeier, J. Brotzge, M. Xue, K. Kloesel, K. Brewster, F. Carr, S. Cruz-Pol, K. Hondl, P. Kollias, 2009: Short-Wavelength Technology and the Potential For Distributed Networks of Small Radar Systems, Bulletin of the American Meteorological Society, Volume 90, Issue 12, 1797-1817.

Melnikov, V. M., D. W. Burgess, D. L. Andra JR., M. P. Foster, J. M. Krause, 2005: K. A. Scharfenberg, D. J. Miller, Terry J. Schuur, P. T. Schlatter, Scott E. Giangrande, 2005: The Joint Polarization Experiment: Polarimetric Radar in Forecasting and Warning Decision Making, Weather and Forecasting, Volume 20, 775-78.

Mercer, A. E., Chad M. Shafer, C. A. Doswell III, Lance M. L., M. B. Richman, 2009: Objective Classification of Tornadic and Nontornadic Severe Weather Outbreaks, Monthly Weather Review, Volume 137, Issue 12 (December 2009) 4355-436.

Mitchell, E. De W., S. V. Vasiloff, G. J. Stumpf, A. Witt, M. D. Eilts, J. T. Johnson, K. W. T., 1998: The National Severe Storms Laboratory Tornado Detection Algorithm, Weather and Forecasting, Volume 13, Issue 2, 352-36.

Moller, A. R., 1978: The Improved NWS Storm Spotters' Training Program at Ft. Worth, Tex., Bulletin of the American Meteorological Society, Volume 59, Issue 12 (December 1978) 1574-1582

Moller, A. R., 2001: Severe Local Storms Forecasting, Meteorological Monographs, Volume 28, Issue 50 (November 2001) 433-480

Moller, A. R., C. A. Doswell III, M. P. Foster, G. R. Woodall, 1994: The Operational Recognition of Supercell Thunderstorm Environments and Storm Structures, Weather and Forecasting, Volume 9, Issue 3, 327-347

Moninger, W. R., C. Lusk, W. F. R. s, J. Bullas, B. de Lorenzis, J. C. McLeod, E. Ellison, J. Flueck, P. D. Lampru, K. C. Young, J. Weaver, R. S. Philips, R. Shaw, T. R. Stewart, S. M. Zubrick, 1991: Shootout-89, A Comparative Evaluation of Knowledge-based Systems That Forecast Severe Weather, Bulletin of the American Meteorological Society, Volume 72, Issue 9 (September 1991) 1339-1354.

Monteverdi, J. P., C. A. Doswell III, G. S. Lipari, 2003: Shear Parameter Thresholds for Forecasting Tornadic Thunderstorms in Northern and Central California, Weather and Forecasting, Volume 18, Issue 2 (April 2003) 357-37.

O'Hora, Fritz and Joan Bech, 2007: Improving weather radar observations using pulse-compression techniques, Meteorological Applications, 14, 389-401.

Pliske, R., D. Klinger, R. Hutton, B. Crandall, B. Knight, and G. Klein, 1997: Understanding skilled weather forecasting: Implications for training and the design of forecasting tools. Contractor Rep. AL/HR-CR-1997-003, Material Command, Armstrong Laboratory, U. S. Air Force, 122

Polger, P. D., B. S. Goldsmith, R. C. Przywarty, J. R. Bocchieri, 1994: National Weather Service Warning Performance Based on the WSR-88D, Bulletin of the American Meteorological Society, Volume 75, Issue 2 (February 1994) 203-21.

Przybylinski, R. W., 1995: The Bow Echo: Observations, Numerical Simulations, and Severe Weather Detection Methods, Weather and Forecasting, Volume 10 (2) 203-21.

Rasmussen. E. N., 2003: Refined Supercell and Tornado Forecast Parameters. Weather and Forecasting 18:3, 530-53.

Rasmussen, E. N., J. M. Straka, R. Davies-Jones, C. A. Doswell III, F. H. Carr, M. D. Eilts, D. R. MacGorman, 1994: Verification of the Origins of Rotation in Tornadoes Experiment: VORTEX, Bulletin of the American Meteorological Society, Volume 75, Issue 6 (June 1994) 995-100.

Ruzanski, E., V. Chandrasekar, Y. Wang, 2011: The CASA Nowcasting System. Journal of Atmospheric and Oceanic Technology, 28, 640-65.

Roberts, Rita D., D. Burgess, M. Meister, 2006: Developing Tools for Nowcasting Storm Severity, Weather and Forecasting, Volume 21, 540-55.

Saffle, R. E., 1976: D/RADEX products and field operation. 17th Conf. on Radar Meteorology, Seattle, AMS, Boston, MA., 555-559.

Schaefer, J. T., 1990: The Critical Success Index as an Indicator of Warning Skill, Weather and Forecasting, Volume 5, 570-57.

Schmeits, M. J., Kees J. Kok, D. H. P. Vogelezang, R. M. van Westrhenen, 2008: Probabilistic Forecasts of (Severe) Thunderstorms for the Purpose of Issuing a Weather Alarm in the Netherlands, Weather and Forecasting, Volume 23, Issue 6 (December 2008) 1253-126.

Schultz, C. J., W. A. Petersen, L. D. Carey, 2011: Lightning and Severe Weather: A Comparison between Total and Cloud-to-Ground Lightning Trends, Weather and Forecasting, Volume 26, Issue 5 (October 2011) 744-75.

Schumacher, R. S., D. T. Lindsey, A. B. Schumacher, J. Braun, S. D. Miller, J. L. Demuth, 2010: Multidisciplinary Analysis of an Unusual Tornado: Meteorology, Climatology, and the Communication and Interpretation of Warnings, Weather and Forecasting, Volume 25, 1412-142.

Serafin, R. J., J. W. Wilson, 2000: Operational Weather Radar in the United States: Progress and Opportunity, Bulletin of the American Meteorological Societ.

Sills, D. M. L., J. W. Wilson, P. I. Joe, D. W. Burgess, R. M. Webb, N. I. Fox, 2004: The 3 November Tornadic Event during Sydney 2000: Storm Evolution and the Role of Low-Level Boundaries, Weather and Forecasting, Volume 19, Issue 1, February 2004) 22-42

Smith, P. L., 1999: Effects of Imperfect Storm Reporting on the Verification of Weather Warnings, Bulletin of the American Meteorological Society, Volume 80, 1099-110.

Stensrud, D. J., L. J. Wicker, K. E. Kelleher, M. Xue, M. P. Foster, J. T. Schaefer, R. S. Schneider, S. G. Benjamin, S. Weygandt, J. T. Ferree, J. P. Tuell, 2009: Convective-Scale Warn-on-Forecast System, Bulletin of the American Meteorological Society, Volume 90, 1487-149.

Stumpf, G. J., A. Witt, E. DeW. Mitchell, P. Spencer, J. T. Johnson, M. D. Eilts, K. W. Thomas, D. W. Burgess, 1998: The National Severe Storms Laboratory Mesocyclone Detection Algorithm for the WSR-88D, Weather and Forecasting, Volume 13, 304-32.

Sun, J., and N. A. Crook, 1994: Wind and thermodynamic retrieval from single-Doppler measurements of a gust front observed during Phoenix-II. Mon. Wea. Rev., 122, 1075-1091.

Sun, J., D. Flicker, and D. K. Lilly, 1991: Recovery of three-dimensional wind and temper¬ature from simulated single-Doppler radar data. J. Atmos. Sci., 48, 876-890.

Uyeda, H. and D. S. Zrnic, 1986: Automatic Detection of Gust Fronts, Journal of Atmospheric and Oceanic Technology, Volume 3, 36-5.

Vasiloff, S. V., 2001: Improving Tornado Warnings with the Federal Aviation Administration's Terminal Doppler Weather Radar, Bulletin of the American Meteorological Society, Volume 82, 861-87.

Wang, H. K., R. Mercer, J. Baron and P. Joe, 2011: Skeleton-based hook echo detection in radar reflectivity data, J. Tech, submitted.

Wang, J. J., T. Keenan, P. Joe, J. Wilson, E. S. T. Lai, F. Liang, Y. Wang, B. Ebert, Q. Ye, J. Bally, A. Seed, M. X. Chen, J. Xue, B. Conway, 2010: Overview of the Beiijing 2008 Olympics Forecast Demonstration Project, China Meteorological Press, 145pp.

Wasula, A. C., L. F. Bosart, K. D. LaPenta, 2002: The Influence of Terrain on the Severe Weather Distribution across Interior Eastern New York and Western New England, Weather and Forecasting, Volume 17, Issue 6 (December 2002) 1277-1289.

Weiss, S. J., C. A. Doswell III, F. P. Ostby, 1980: Comments on Automated 12-36 Hour Probability Forecasts of Thunderstorms and Severe Local Storms, Journal of Applied MeteorologyVolume 19, Issue 11, 1328-1333.

Weisman, M. L., 2001: Bow Echoes: A Tribute to T. T. Fujita, Bulletin of the American Meteorological Society, Volume 82, 97-116

Weisman, M. L., J. B. Klemp, 1984: The Structure and Classification of Numerically Simulated Convective Stormsin Directionally Varying Wind Shears, Monthly Weather Review, Volume 112, Issue 12 (December 1984) 2479-2498

Weisman, M. L., R. Rotunno, 2000: The Use of Vertical Wind Shear versus Helicity in Interpreting Supercell Dynamics, Journal of the Atmospheric Sciences, Volume 57, Issue 9 (May 2000) 1452-1472

Weisman, M. L., R. Rotunno, 2004: "A Theory for Strong Long-Lived Squall Lines" Revisited, Journal of the Atmospheric Sciences, Volume 61, Issue 4 (February 2004) 361-382

Westefeld, J. S., A. R. Less, T. Ansley, H.S. Yi, 2006: Severe-Weather Phobia, Bulletin of the American Meteorological Society, 2006: Volume 87, 747-74.

Wilson, J. W., E. A. Brandes, 1979: Radar Measurement of Rainfall — A Summary, Bulletin of the American Meteorological Society, Volume 60, 1048-105.

Wilson, J., R. Carbone, H. Baynton, R. Serafin, 1980: Operational Application of Meteorological Doppler Radar, Bulletin of the American Meteorological Society, Volume 61, 1154-116.

Wilson, J. W., N. A. Crook, C. K. Mueller, J. Sun, M. Dixon, 1998: Nowcasting Thunderstorms: A Status Report, Bulletin of the American Meteorological Society, Volume 79, 2079-209.

Wilson, J. W., E. E. Ebert, T. R. Saxen, R. D. Robertss, C. K. Mueller, M. Sleigh, C. E. Pierce, A. Seed, 2004: Sydney 2000 Forecast Demonstration Project: Convective Storm Nowcasting, Weather and Forecasting, Volume 19, 131-15.

Wilson, J. W., Y. Feng, Min Chen, Rita D. R. s, 2010: Nowcasting Challenges during the Beijing Olympics: Successes, Failures, and Implications for Future Nowcasting Systems, Weather and Forecasting, Volume 25, 1691-171.

Wilson, J. W., J. A. Moore, G. B. Foote, B. Martner, T. Uttal, J. M. Wilczak, A. R. Rodi, 1988: Convection Initiation and Downburst Experiment (CINDE), Bulletin of the American Meteorological Society, Volume 69, 1328-134.

Wilson, J. W., R. M. Wakimoto, 2001: The Discovery of the Downburst: T. T. Fujita's Contribution, Bulletin of the American Meteorological Society, Volume 82, 49-6.

Winston, H. A., 1998: A Comparison of Three Radar-Based Severe-Storm-Detection Algorithms on Colorado High Plains Thunderstorms, Weather and Forecasting, Volume 3, 131-140.

Winston, H. A., L. J. Ruthi, 1986: Evaluation of RADAP II Severe-Storm-Detection Algorithms, Bulletin of the American Meteorological Society, Volume 67, 145-15.

Witt, A., M. D. Eilts, G. J. Stumpf, J. T. Johnson, E. De W. Mitchell, K. W. T., 1998a: An Enhanced Hail Detection Algorithm for the WSR-88D, Weather and Forecasting, Volume 13, 286-30.

Witt, A., M. D. Eilts, G. J. Stumpf, E. De W. Mitchell, J. T. Johnson, K. W. T., 1998b: Evaluating the Performance of WSR-88D Severe Storm Detection Algorithms, Weather and Forecasting, Volume 13, 513-51.

WMO, 2008: Commission on Instruments, Methods and Observations Guide, Chapter 9, 7th edition, available at
http://www.wmo.int/pages/prog/www/IMOP/publications/CIMO-Guide/CIMO%20Guide%207th%20Edition,%202008/Part%20II/Chapter%209.pdf

Wong, W. K., L. H. Y. Yeung, Y. C. Wang & M. Chen, 2009: Towards the Blending of NWP with Nowcast — Operation Experience in B08FDP, WMO Symposium on Nowcasting, 30 Aug-4 Sep 2009, Whistler, B. C., Canada.

Wong K. Y., C. L. Yip and P. W. Li, 2007: A Novel Algorithm for Automatic Tropical Cyclone Eye Fix Using Doppler Weather Radar, Meteorological Applications, 14, 49-5.

Yeung, L. H. Y., E. S. T. Lai & P. K. Y. Chan, 2008: Thunderstorm Downburst and Radar-based Nowcasting of Squalls, the Fifth European Conference on Radar in Meteorology and Hydrology, Helsinki, Finland 30 June - 4 July 200.

Yeung, L. H. Y., W. K. Wong, P. K. Y. Chan & E. S. T. Lai, 2009: Applications of the Hong Kong Observatory nowcasting system SWIRLS-2 in support of the 2008 Beijing Olympic Games. WMO Symposium on Nowcasting, 30 Aug-4 Sep 2009, Whistler, B. C., Canada.

Yeung, L. H. Y., E. S. T. Lai & S. K. S. Chiu, 2007: Lightning Initiation and Intensity Nowcasting Based on Isothermal Radar Reflectivity – A Conceptual Model, the 33rd International Conference on Radar Meteorology, Cairns, Australia, 6-10 August 2007.

Zipser, E. J., C. Liu, D. J. Cecil, S. W. Nesbitt, D. P. Yorty, 2006: Where are the most intense thunderstorms on Earth?, BAMS, 87 (8), 1057-1071.

Zrnic, D. S., D. W. Burgess, L. D. Hennington, 1985: Automatic Detection of Mesocyclonic Shear with Doppler Radar, Journal of Atmospheric and Oceanic Technology, Volume 2, 425-438.

3

Aviation Applications of Doppler Radars in the Alerting of Windshear and Turbulence

P.W. Chan[1] and Pengfei Zhang[2]
[1]Hong Kong Observatory, Hong Kong,
[2]University of Oklahoma, Norman, OK,
[1]China
[2]USA

1. Introduction

Doppler radars are indispensable nowadays in the assurance of aviation safety. In particular, many airports in the world are equipped with Terminal Doppler Weather Radar (TDWR) in the alerting of low-level windshear and turbulence. The microburst alerts from certain TDWR are taken as "sky truth" and the aircraft may not fly when microburst alerts are in force.

This chapter summarizes some recent developments on the aviation applications of TDWR in Hong Kong (Figure 1). It first starts with a case study of a typical event of microburst alert associated with severe thunderstorms. The applications of TDWR in the alerting of windshear and turbulence are then described, namely, in the calculation of windshear hazard factor using the radial velocity data from the radar, and the calculation of eddy dissipation rate based on the spectrum width data of the radar. It is hoped that this chapter could serve as an introduction to the aviation applications of TDWR, for the reference of the weather services of other airports.

2. A typical microburst event leading to missed approaches of aircraft

The missed approaches at the Hong Kong International Airport (HKIA) took place during the overnight period of 8 to 9 September 2010 when intense thunderstorm activity brought heavy rain and frequent lightning to the whole Hong Kong. During the period, an intense rain band with north-south orientation swept from east to west across Hong Kong. More than 50 millimeters of rainfall in an hour were generally recorded over the territory and a record-breaking number of 13,102 cloud-to-ground lightning strokes were registered during the hour just after midnight. When the thunderstorms edged close to the HKIA which is situated at the western part of the territory, gusty strong easterlies from the downdraft of the thunderstorm first affected the flight paths east of the airport resulting in an abrupt change in the prevailing winds from southwesterlies to easterlies.

Two flights, which tried to land as the thunderstorms approached HKIA, aborted landing and diverted to Macao eventually. Both flights approached the HKIA from the east under the prevailing southwesterly winds (Figure 2). The first aircraft went around twice. The first

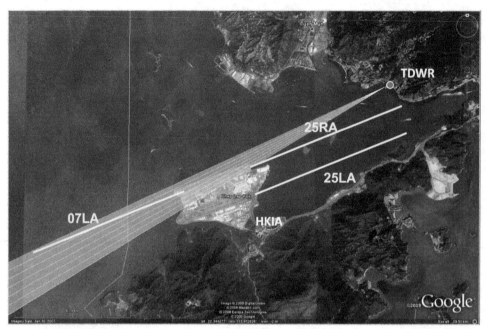

Fig. 1. The locations of the Hong Kong TDWR (red dot) radar and Hong Kong International Airport (HKIA). The blue beams illustrate the radar beams over the runways corridor 07LA of the airport with 1° azimuth interval. Three yellow lines indicate the approach paths and their names are marked.

Fig. 2. Flight paths of the two aircraft which had to conduct missed approach. Red line indicated the flight path for the first aircraft and yellow for the second aircraft. Orange wind barbs showed the locations of aircraft when tailwind was encountered. The 1st and 2nd aircraft recorded tailwind of 37 and 22 knots respectively.

aborted landing was due to technical consideration. In the second approach at around 00:08 HKT (=UTC + 8 hours), it encountered strong tailwind. Landing was subsequently aborted and the aircraft diverted to Macao thereafter. Four minutes later, the second aircraft followed the same glide path of the first aircraft but also failed to land at the HKIA because of the same reason, i.e. the strong tailwind. The aircraft was also diverted to Macao at 00:12 HKT.

Flight data retrieved from the flight data recorders of the two aircraft was analyzed to reveal the meteorological conditions encountered by aircrafts. It appeared that the missed approach was attributable to the strong tailwind which exceeded the airline pre-defined threshold, namely 15 knots for tailwind landing.

According to the flight data, the first aircraft experienced more than 15 knots tailwind after it descended to below 1600 feet (Figure 3(a)) in its second approach. The tailwind increased from 25 knots when the aircraft descended to 780 feet (labeled 'A' in Figure 3(a)) and strengthened to 37 knots at 708 feet at 00:08 HKT (labeled 'B' in Figure 3(a)), which far exceed the limit for tailwind landing. As a result, diversion to other airport was conducted.

The second aircraft also experienced the tailwind of around 15 knots when it descended to around 1600 feet. The tailwind increased and reached 19 knots when the aircraft descended to 1423 feet (labeled 'C' in Figure 3(b)) but then decreased and fluctuated between 7 to 12 knots when the aircraft further descended to 1028 feet (labeled 'D' in Figure 3(b)). At around 00:12 HKT, the tailwind started to strengthen again and exceeded 15 knots. The maximum tailwind experienced by the aircraft was 22 knots, which also exceeded the limit for tailwind landing, at 859 feet above the runway (labeled 'E' in Figure 3(b)). Similar to the first aircraft, the second aircraft executed a missed approach due to the strong tailwind and was diverted to Macao.

The TDWR also captured the wind conditions when the two aircraft conducted missed approaches. Figures 4(a) and 4(b) showed the radial velocity measured by TDWR at 0008 HKT and 0012 HKT 9 September respectively. Gusts reaching 27 m/s (i.e. around 50 knots) were captured by the TDWR over the eastern part of the HKIA. The zero isotach, which marked the leading edge of the shear line, agreed well with that identified based on anemometer data.

The HKO Windshear and Turbulence Alerting System (WTWS) integrates windshear and turbulence alerts generated by different algorithms such as Anemometer-based Windshear Alerting Rules-Enhanced (AWARE) (Lee, 2004), LIDAR Windshear Alerting System (LIWAS) (Shun and Chan, 2008), TDWR alerts and other algorithms. Alerts are then generated for 8 runway corridors (north runway and south runway have two arrival and two departure corridors each) and shown on a graphical display, the WTWS display.

At 0008 HKT, the zero isotach over the HKIA detected by the TDWR was analyzed as a gust front and was shown on the WTWS display (Figure 5(a)). In addition, microburst alerts, which represent windshear loss of 30 knots or more with precipitation, were provided by TDWR to the east of the HKIA; windshear alerts were generated from AWARE over the runways; turbulence alerts were in force due to the thunderstorm to the north of the HKIA. Over the 8 corridors of the HKIA, all had windshear alerts with magnitude ranging from +25 to +30 knots. At 0012 HKT, although the gust front was not detected by the TDWR

(Figure 5(b)) any more, using the surface anemometers and TDWR base data, windshear alerts with magnitude ranging from +15 to +25 knots were issued for the four western corridors. Meanwhile, areas with the microburst alerts shifted westwards and affected the eastern corridors. WTWS issued microburst alerts of -35 knots to the four eastern corridors. During the event, the WTWS functioned properly and was able to provide adequate warning to the aircraft of the windshear to be expected due to the thundery weather.

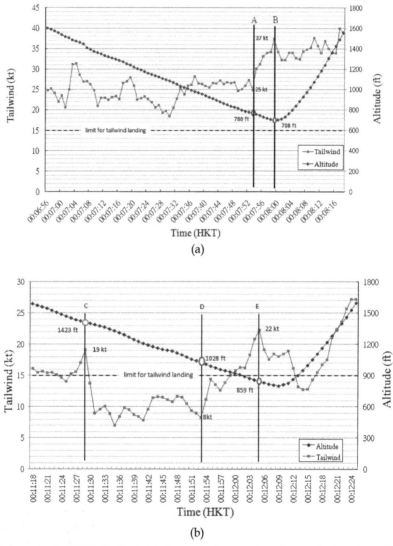

Fig. 3. Time series in HKT of tailwind in knots (red square) and aircraft altitude in feet (blue diamond) retrieved from the flight data recorders. (a) Flight data for the first aircraft. Tailwind reached 37 knots at 00:08 HKT. (b) Flight data for the second aircraft.

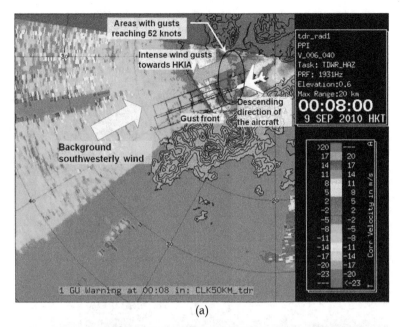

(a)

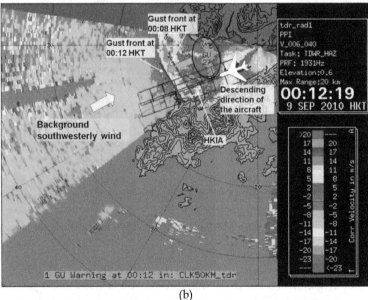

(b)

Fig. 4. Velocity measured by TDWR on 9 September 2010. The cool/warm colors represent winds towards/away from the TDWR. Area with gusts reaching 27 m/s was circled in black. The zero isotach (gust front) was in purple. (a) TDWR image at 0008 HKT; (b) TDWR images at 0012 HKT. The zero isotach (gust front) moved westwards to the western end of HKIA.

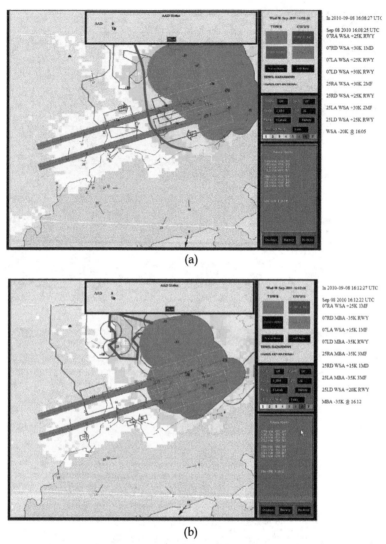

(a)

(b)

Fig. 5. WTWS display on 9 September 2010. Gust front analyzed by TDWR (purple line) over the HKIA; microburst alerts generated by TDWR (red solid band-aids); windshear alerts generated by AWARE (red hollow rectangles), by TDWR (red hollow irregular polygons); by LIDAR (red arrows, over the runways only); turbulence alert generated by TDWR (brown polygon with dots). Black numbers were the windshear magnitude in knots. (a) 0008 HKT on 9 September 2010. A gust front was over the HKIA. Windshear alerts were issued by the WTWS for all runway corridors. LIDAR data was highly attenuated by precipitation and could only detect windshear over the runway. (b) 0012 HKT on 9 September 2010. Microburst alerts of -35 knots were issued to the four eastern corridors. Windshear alerts with magnitude ranging from +15 to +25 knots were issued for the four western corridors.

3. Windshsear hazard factor based on TDWR

In aviation meteorology, windshear refers to a sustained change of wind speed and/or wind direction that causes the aircraft to deviate from the intended flight path. Low-level windshear (below 1600 feet) could be hazardous to the arriving/departing aircraft. Hong Kong is situated in a subtropical coastal area and it is common to have intense convective weather in the spring and summer. To alert low-level windshear associated with microburst and gust front, a TDWR is operated by the Hong Kong Observatory (HKO) in the vicinity of HKIA (Figure 1). It is a C-band radar with 0.5-degree half-power beam width scanning over the airport and determines convergence/divergence features along the runway orientation from the Doppler velocities. Windshear alerts are generated when the velocity change is 15 knots or more.

Another index that quantifies the windshear threat is the F-factor (Proctor et al., 2000). It is based on the fundamentals of flight mechanics and the understanding of windshear phenomena. The F-factor could also be calculated from the Quick Access Recorder (QAR) data recorded on the commercial jets (Haverdings, 2000). In this study, an attempt is made to calculate F-factor for some typical microburst events at HKIA based on the TDWR measurements and the results are compared with the F-factor determined from the QAR data.

F-factor is calculated from TDWR's radial velocity data in two steps. First of all, convergence/divergence features are identified from the TDWR data. Then F-factor is determined from each convergence/divergence feature by assuming a wind field model of microburst. The two steps are briefly described below.

To compute convergence/divergence features, the method described in Merritt (1987) is adopted. The TDWR microburst detection algorithm identifies microburst by searching for significant velocity difference along a radial in a search window of 4 range gates (4 x 150 metres per gate = 600 metres in length, and one degree in azimuth). If the windshear along a search window is divergent (i.e. radial wind generally increases with increasing distance from the radar), the search window is taken to be a divergence shear segment. Likewise, convergence shear segment is also identified.

Two divergence/convergence segments are associated as a divergence/convergence shear features if their minimum overlap in range is 0.5 km or if their maximum angular spacing is 2 degrees azimuth. A divergence/convergence region contains at least 4 shear segments with a minimum length of 0.95 km and a minimum area of 1 km². Moreover, the maximum velocity difference among the shear segments inside a divergence region should be at least 5 m/s. As such, the shear within a divergence region is at least 5 m/s per 600 m, i.e. 0.008 m/s/m.

F-factor is related to the total aircraft energy and its rate of change, and is defined to be:

$$F = \frac{\dot{W_x}}{g} - \frac{w}{V_a} \tag{1}$$

where W_x is the component of atmospheric wind directed horizontally along the flight path (direction x) and $\dot{W_x}$ its rate of change, g the acceleration due to gravity, w the updraft of the atmosphere, and V_a the airspeed of the aircraft. By estimating the updraft from mass continuity constraint, it is shown to be equivalent to:

$$F = \frac{\partial W_x}{\partial x}\left[\frac{V_g}{g} + \frac{2h}{V_a}\right] \tag{2}$$

where V_g is the ground speed of the aircraft, and h the altitude above ground.

For each convergence/divergence feature captured by the TDWR, the velocity change ΔU and the distance over which this change occurs ΔR are calculated. It is shown in Hinton (1993) with reference to a microburst model that F-factor could be calculated from:

$$F = K\frac{\Delta U}{\Delta R}\left[\left(\frac{\Delta R}{L}\right)^2 - \left(\frac{\Delta R}{L}\right)^3 \frac{\sqrt{\pi}}{2\alpha} erf\left(\frac{\alpha L}{\Delta R}\right)\right] \bullet \left[\frac{V_g}{g} + \frac{2h_r}{V_a}\right] \tag{3}$$

where $K = 4.1925$, $a = 1.1212$, h_r the above-ground-level (AGL) altitude of the TDWR radar beam, L the characteristic shear length of 1000 m, and $erf(y)$ the error function.

The microburst model in Hinton (1993) includes a shaping function which describes the change in microburst outflow with altitude. This function is given by:

$$p(h) = \frac{e^{-0.22h/H} - e^{-2.75h/H}}{0.7386} \tag{4}$$

where h is the altitude above ground and H the altitude of maximum outflow speed (assumed to be 90 m). The F-factor F_1 from the TDWR at the radar beam altitude h_1 is then related to the F-factor F_2 of the aircraft at the altitude h_2 by the following equation:

$$F_2 = F_1 \frac{p(h_2)\left(\dfrac{V_g}{g} + \dfrac{2h_2}{V_a}\right)}{p(h_1)\left(\dfrac{V_g}{g} + \dfrac{2h_1}{V_a}\right)} \tag{5}$$

Combining (3) – (5) and with ΔU and ΔR determined, the F-factor associated with a divergence/convergence feature at the altitude of the aircraft along the glide path could be calculated.

For the formulation in (1), F-factor is positive if the windshear is performance decreasing (headwind decreasing or downdraft) and negative if the windshear is performance increasing (headwind increasing or updraft). As discussed in Proctor et al. (2000), for onboard windshear systems, the windshear is considered to be hazardous if F is greater than 0.1, and a *must alert* threshold is set to be 0.13. The must alert threshold means a wind shear alert must be issued when that threshold is reached/exceeded.

A microburst event that affected HKIA on 18 May 2007 is considered here as an illustration of the method. In the evening of that day, a surface trough of low pressure lingered around the south China coast, bringing unsettled weather to the region. Between 09 and 10 UTC (5 and 6 p.m. of 18 May 2007), a band of strong radar echoes with east-northeast to west-southwest orientation moved southeastwards from inland areas across the coast. At HKIA, the TDWR issued microburst alerts of 30 knots headwind loss for the aircraft between 09:20 and 09:27 UTC.

Figure 6(a) shows the moment when a microburst associated with the thunderstorms affected the runway corridors to the east of HKIA. Divergent flow feature was found at 0.6-degree conical scan of TDWR. For an aircraft arriving at the north runway of HKIA (location in Figure 1) from the east, the windshear associated with the microburst is performance decreasing (due to decreasing headwind). Using the formulae above, the F-factor for the microburst is determined to be about 0.14, which exceeds the must alert threshold and the windshear associated with the microburst is considered to be hazardous to the aircraft. Flight data are obtained for an aircraft arriving at the north runway from the east at that time. They are processed by the algorithm in Haverdings (2000) and the variation of F-factor along the glide path is shown in Figure 6(b). At about the location of the microburst (near the eastern threshold of the north runway), the F-factor is found to be about 0.13, which is generally consistent with the value determined from TDWR data. Thus for microburst associated with the thunderstorm, the F-factor determined from TDWR measurements and that from QAR data of the aircraft are comparable with each other. The other peaks/troughs of F-factor from the QAR data (Figure 6(b)) are not revealed in the TDWR measurements. They may not be properly handled by the microburst model for F-factor calculation.

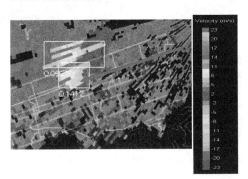 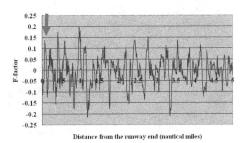

Fig. 6. (a) Divergence features (highlighted in lighter colours) associated with microburst on 18 May 2007, overlaid on the radial velocity from the TDWR (colour scale on the right). F-factor of each feature is given as a number next to the box indicating the location of the feature. (b) F-factor as recorded on an aircraft flying at about the same time as in (a) along the glide path shown as a red arrow in (a). The red arrow in (b) is the approximate location of the windshear feature encountered by the aircraft.

To study the change in the F-factor following the evolution of the microburst, the intense convective event on 8 June 2007 is considered. Severe gusts associated with thunderstorms and microburst with a recorded maximum of 35.9 m/s affected HKIA in the morning of that day. A helicopter parked on the apron toppled in strong winds during the passage of the intense storm cells. We just focus on the windshear hazard associated with the microburst. The divergence features determined from the radial velocity of the TDWR at 0.6-degree conical scans are shown in Figure 7. Stronger winds associated with the microburst got closer to the ground level (about 260 m above mean sea level at the location of the microburst) in a short time interval within 3 minutes, with the maximum value of towards-the-radar velocity increasing from 18 m/s (dark blue in Figure 7) to 23 m/s (magenta in Figure 7). As a result, the F-factor increases in magnitude from 0.14 to 0.23, which exceeds the must alert threshold. The TDWR-based F-factor provides a good indication about the level of hazard associated with an evolving microburst.

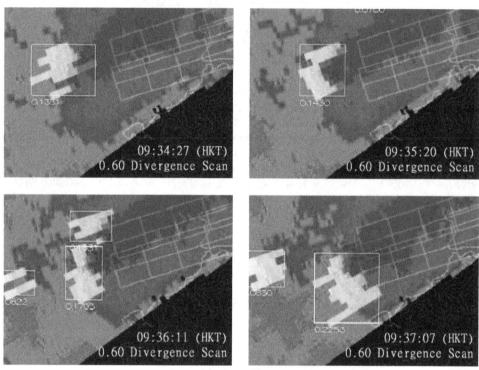

Fig. 7. Time series of the divergence feature associated with a microburst on 8 June 2007. The feature is highlighted in lighter colour and enclosed in a box. The number next to the box is the F-factor calculated for the feature. The background is the radial velocity from the TDWR, with the colour scale given in Figure 6.

Besides intense convective weather, the windshear hazard in terrain-disrupted airflow is also studied. The Typhoon Prapiroon case on 3 August 2006 is considered. On that day, Prapiroon was located at about 200 km to the southwest of Hong Kong over the South China Sea and tracked northwest towards the western coast of southern China. This typhoon brought about gale-force east to southeasterly airflow to Hong Kong. Due to complex terrain to the south of the airport, airflow disturbances occurred inside and around HKIA. Divergent flow features were observed near the airport from time to time. Figure 8(a) shows such a feature at 0.6-degree conical scan of the TDWR at about 4:47 a.m., 3 August. The F-factor associated with this feature is about 0.22, which exceeds the must alert threshold for windshear. An aircraft landed at the north runway of HKIA from the west at about that time (within one minute). The variation of the F-factor determined from QAR data along the glide path is given in Figure 8(b). At the location of the microburst, the F-factor from the aircraft is comparable with that calculated from the TDWR data, even for this case of terrain-disrupted airflow. As discussed in the first case study, the other peaks/troughs of F-factor from the QAR data (Figure 8(b)) are not revealed in the TDWR measurements. They may not be properly handled by the microburst model for F-factor calculation.

 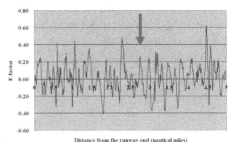

Fig. 8. (a) Divergence features (highlighted in lighter colours) associated with windshear in terrain-disrupted airflow on 3 August 2006, overlaid on the radial velocity from the TDWR (colour scale given in Figure 6). F-factor of each feature is given as a number next to the box indicating the location of the feature. (b) F-factor as recorded on an aircraft flying at about the same time as in (a) along the glide path shown as a blue arrow in (a). The red arrow in (b) is the approximate location of the windshear feature encountered by the aircraft.

4. Calculation of turbulence intensity

The measurement of spectrum width is determined not only by the Doppler velocity distribution and density distribution of the scatterers within the resolution volume, but also radar observation parameters like beamwidth, pulse width, antenna rotation rate, etc. According to Doviak and Zrnic (2006), there are five major spectral broadening mechanisms that contribute to the spectrum width measurements, which can be written as follow

$$\sigma_v^2 = \sigma_s^2 + \sigma_t^2 + \sigma_\alpha^2 + \sigma_d^2 + \sigma_o^2 \qquad (6)$$

where σ_s represents mean wind shear contribution, σ_t represents turbulence, σ_α represents antenna motion, σ_d represents different terminal velocities of hydrometeors of different sizes, and σ_o represents variations of orientations and vibrations of hydrometeors. Except σ_s and σ_t, the rest of the terms on the right hand side of the Eq.(6) are considered to be negligible for the measurements of σ_v in this paper (Brewster and Zrnic, 1986). Thus the turbulence term σ_s can be obtained,

$$\sigma_t^2 = \sigma_v^2 - \sigma_s^2. \qquad (7)$$

In the Eq.(7), mean wind shear width term σ_s can be decomposed into three terms due to mean radial velocity shear at three orthogonal directions in radar coordinate(Doviak and Zrnic, 2006):

$$\sigma_s^2 = \sigma_{s\theta}^2 + \sigma_{s\phi}^2 + \sigma_{sr}^2 = (r_0 \sigma_\theta k_\theta)^2 + (r_0 \sigma_\phi k_\phi)^2 + (\sigma_r k_r)^2 , \tag{8}$$

where $\sigma_r^2 = (0.35c\tau/2)^2$, $\sigma_\theta^2 = \theta_1^2/16\ln2$, and $\sigma_\phi^2 = \theta_1^2/16\ln2$. Here $c\tau/2$ is range resolution, and θ_1 is the one-way angular resolution (i.e., beamwidth). k_θ, k_ϕ, and k_r are the components of shear along the three orthogonal directions.

In order to use σ_t to estimate eddy dissipation rate (EDR) ε, it must be assumed that within radar resolution volume turbulence is isotropic and its outer scale is larger than the maximum dimension of the radar's resolution volume (which is indicated as V_6). Under these assumptions, in the case of $\sigma_r \leq r\sigma_\theta$ the relation between turbulence spectrum width σ_t and EDR ε can be approximately written as (Labitt, 1981)

$$\varepsilon \approx \frac{0.72\sigma_t^3}{r\sigma_\theta A^{3/2}} , \tag{9}$$

where A is constant (i.e., about 1.6). When $\sigma_r \geq r\sigma_\theta$, the relation can be approximated by

$$\varepsilon \approx \left[\frac{<\sigma_t^2>^{3/2}}{\sigma_r(1.35A)^{3/2}}\right]\left(\frac{11}{15} + \frac{4}{15}\frac{r^2\sigma_\theta^2}{\sigma_r^2}\right)^{-3/2} \tag{10}$$

Eqs. (9) and (10) are used to estimate EDR using Hong Kong TDWR observed spectrum width.

In hazardous weather mode, the Hong Kong TDWR conducts sector scans from azimuth 182° to 282° (i.e., confined to the approach and departure paths). Each sector scan takes about 4 minutes. Thus, the low altitude wind shear can be detected within a minute. The range and angular resolutions of the radar are 150 m and 0.5° respectively. The maximum range reaches 90 km. The radar data includes reflectivity, Doppler velocity, spectrum width, and signal-to-noise ratio (SNR) recorded with the azimuth interval of 1°.

Based on the Eqs. (9) and (10), EDR can be estimated when spectrum width observation is available. In this feasibility study, EDR estimation is only performed at the lowest elevation angle of 0.6°. The vertical wind shear contribution to the EDR is calculated by using spatially averaged mean Doppler velocity at two lowest elevation angles. Because the closest two elevation angles at lowest level are 0.6° and 1.0° at scans 11 and 12, vertical wind shear is calculated by using the Doppler velocity fields at these two scans. For simplicity, EDR is estimated at scan 17 with elevation angle of 0.6°. Azimuthal and radial wind shear is also calculated at this scan. So in the current algorithm, one EDR field at elevation angle of 0.6° will be generated for each volume scan.

The control of the TDWR spectrum width data quality is very important for EDR estimation. It has been found that there is a variety of sources of errors in spectrum width measurements in previous studies (Fang et al. 2004). Especially if signal to noise ratio (SNR) is low, spectrum width measurements have large variance. In this study, SNR > 20 dB is assigned as a simple and straightforward threshold for the EDR estimates. In other words, EDR at the gate with SNR < 20 dB is marked as missing data (MD) in our algorithm. In the future, more comprehensive quality control processor will be designed and implemented in our algorithm to deal with other error sources.

Following international practice, EDR values are classified into four categories in terms of the intensity of turbulence. For convenience and in line with alerting purpose of low-level turbulence, EDRs in the following figures and context will be labeled or indicated as insignificant (LL), light (L), moderate (M), and severe (S) instead of its value. It is also worth mentioning that EDR values presented in this paper are derived from the spectrum width data after smoothing by using a 9 point median filter along the radar beams in order to suppress the fluctuations in the determination of spectrum width values.   This kind of fluctuation is expected, for instance, to arise from the limited and finite number of data points in the digitization of the spectrum of the return signal.

The spectrum width errors are large in region of low SNR. Here we selected a case to demonstrate the importance of the SNR threshold in the quality control of EDR data. Around 21 UTC on 6 June 2008, the TDWR radar observed thunderstorms over HKIA. Without SNR threshold, estimated EDR suggested severe turbulence region (red color; Figure 9(a)) in the region about azimuth of 270º and centered at about 25 km. High spectrum widths (~4.5 m/s) are indeed measured in this region (see Figure 9(c)). But reflectivity (Figure 9(e)) and SNR (Figure 9(d)) are around -8 dBZ and 10 dB respectively. The relatively large spectrum widths in this region can be caused by incorrect noise power estimates (Fang et al., 2004). To avoid such biases, we use a SNR threshold of 20 dB as recommended by Fang et al., (2004).

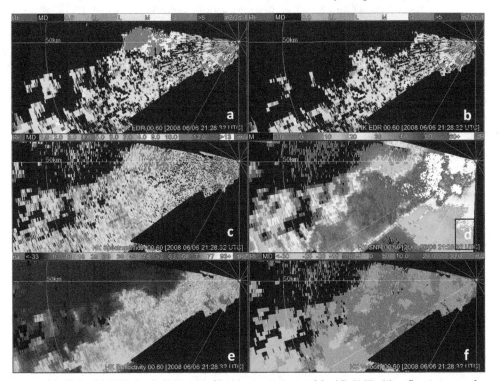

Fig. 9. (a) EDR, (b) EDR with SNR> 20 dB, (c) spectrum width, (d) SNR, (e) reflectivity, and (f) Doppler velocity at elevation angle of 0.6º at 21:28 UTC on 6 June 2008. Range ring is 50 km and azimuths are every 30º.

On the other hand, there are two small regions near the radar at the range of 6 km where EDR is also high. But in this region there is relatively strong horizontal shear of the radial wind component (Figure 9(f); green color identifies the wind has a component toward the radar and red color indicates wind is away from the radar). Furthermore, the reflectivity is about 10 dBZ and SNR is around 35 dB. Because this region is on the downwind side of Lantau Island, the ambient flow (green in Figure 9(f)) is blocked by the Island and back flow (red in Figure 9(f)) is induced. The wind shear contributions, computed using Eq. (8), have been removed from the calculation EDR presented in Figure 9(a). Thus the EDR should not be biased by strong shear of mean radial wind. Thunderstorm outflow may be another reason for the severe turbulence in this region. Because there is no strong horizontal shear of the Doppler velocity field in the region 270° and 25 km, we conclude that the large EDRs presented in in that region of Figure 9(a) are unrealistic. After a threshold SNR> 20 dB is applied, it can be seen that these large EDR values are removed (Figure 9(b)).

Using the Hong Kong TDWR observations in 2006 and 2008, many EDR maps were produced and examined. Here wind shear contribution has been removed from spectrum width measurements. Here the mean wind shears in horizontal and vertical directions are calculated by using mean radial velocity field smoothed by a 9 points median filter along the radar beam in the Eq.(8). Figure 10 shows two typical EDR maps during light rain at 21:32 UTC on 27 April 2006 (Figure 10(a)) and during a thunderstorm at 13:17 UTC on 13 June 2008 (Figure 10(b)). For most of the scanned area, EDR is low and turbulence is classified as insignificant or light (green and light blue). Small pockets of moderate and severe turbulence (yellow and red) are scattered in the scanned area. Near the Lantau Island, moderate and severe levels of turbulence are frequently observed in the cases we studied. The blockage of the Island on the ambient flow may be a reason for the occurrence of the turbulent airflow. Based on the numerical simulations, Clark et al. (1997) and Chan (2009) found that mechanical effect of a mountainous island is a source of the generation of the turbulence.

Clear air cases have been investigated as well, but we found that SNR of the Hong Kong TDWR is too low to provide reliable and meaningful EDR maps.

After the EDR maps were generated, EDR profiles along the flight paths can be compared with aircraft measured EDR. A total of 14 cases are selected to make the comparison. The aircraft EDRs are estimated based on the vertical wind measured by aircraft (Cornman et al., 2004).

Radar derived EDR profile is constructed by selecting the EDR in a resolution volume V_6 closest to the flight path and at an elevation angle of 0.6°. There are still differences in the measurement heights between the aircraft and the radar beam for these two EDR datasets. Only a part of the flight path is covered by the radar beam. For example, aircraft approaching runway 25RA is in the radar beam only at the distance between 0.5 and 1.5 nm from the end of runway. From this point of view, EDRs estimated by aircraft and the radar would be compared within this distance interval. It should also be mentioned that radar estimated EDR is based on the spectrum width of the Doppler velocity, i.e. velocity in the radial direction along a radar beam. On the other hand, the aircraft estimated EDR is based on the vertical wind. As such, the two EDR datasets are derived from different components of the wind. Put aside errors in measurement, in order to have agreement turbulence must be isotropic.

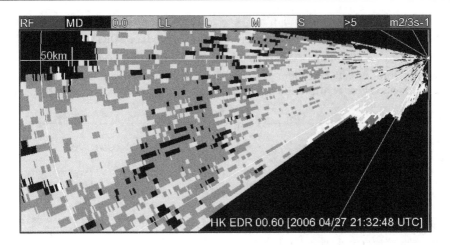

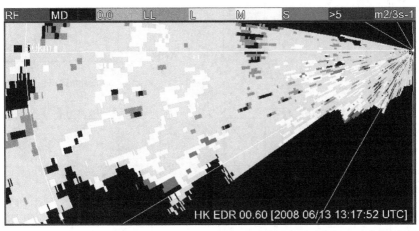

Fig. 10. EDR maps (a) at 21:32 UTC on 27 April 2006 and (b) 13:17 UTC on 13 June 2008. The mountainous Lantau Island is located to the south of the radar scans.

Another issue of the comparison is the contribution of mean wind shear to the measured spectrum width. For the estimation of EDR, the contribution of wind shear has to be extracted from the radar measured spectrum width. But for the comparison with aircraft measured EDR or even turbulence alert for aviation safety, wind shear might not need to be removed. For example if the aircraft experiences a sharp change in altitude, this may not be caused by isotropic turbulence but it is a measure of aircraft response to vertical shear of mean wind. As such, the aircraft estimated EDR based on vertical velocity may be slightly higher. Pilots and passengers in aircraft may also experience severe "turbulence", which is a combination of the effects of both turbulence and wind shear.

Scatterplots of median and maximum EDR along the 5 nm of flight paths estimated by aircraft and radar are shown in Figure 11. Two plots for each are shown; one in which mean wind shear contributions to the observed spectrum widths are removed and a second plot in which mean wind shear contribution has been retained. All median EDRs are smaller than $0.4\,m^{2/3}/s$ (i.e., moderate or light turbulence). 13 of 14 median EDRs indicate turbulences are light. Based on maximum EDRs, two severe turbulent patches (EDR > $0.5\,m^{2/3}/s$) are detected by both aircraft and radar with wind shear, but they are not on the same flight paths. With wind shear contribution, median and maximum radar EDRs evidently increase.

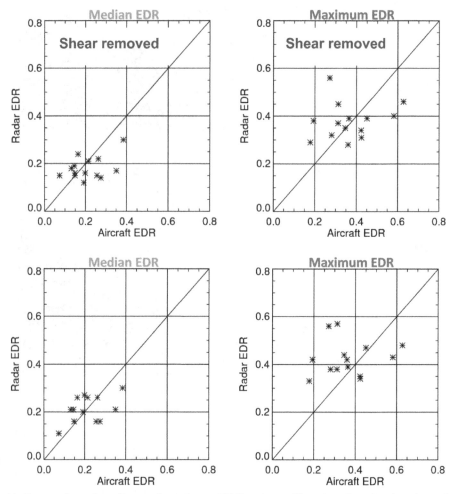

Fig. 11. Scatterplots of median and maximum EDR estimated by aircraft and radar along the 5 nm of flight paths for the selected 14 cases.

Comparing maximum intensity between aircraft and radar without wind shear, 8 of the 14 cases are in the same category. Seven of them are moderate turbulence. For 4 aircraft estimated light turbulence cases, the radar tends to overestimate them as moderate (3 cases) and severe (1 case) with wind shear contribution. After closer examination of the overestimation case at 07:17 UTC on 25 June 2008, it is found that the maximum severe turbulence only occurs at one radar gate at the distance of 0 nm, closest to the end of the runway. It is noted that at this location, the radar beam is higher than the flight path by about 160m.

We have also compared aircraft and radar estimated EDR profiles including wind shear contribution along the aircraft flight path. For this case, aircraft B777 flew through a storm with maximum reflectivity of 42 dBZ and landed in clouds and light rain at HKIA.

Figure 12 shows the EDR estimated by aircraft and the radar along the flight path 25RA around 13:05 UTC on 19 April 2008. It is one of the two cases in which severe turbulence was encountered by the aircraft. Blue dots in Figure 13 represent the EDR estimated by the aircraft as it was landing at HKIA. Three peaks over 0.5 $m^{2/3}/s$, classified as severe turbulence, are recorded at distance of 0.77, 3.65, and 4.90 nm away from the runway end. EDR profiles estimated by using radar data at an elevation angle of 0.6° with the wind shear contribution included in the volume scans around 13:05 UTC are overlaid onto the aircraft estimated EDR in Figure 13. The radar estimated EDR profiles at 13:01, 13:05, and 13:09 UTC (brown dots, red squares, and green dots in Figure 13) matches well with aircraft EDR between distance of 0.5 and 1.5 nm, shaded in green color in Figure 13, where the aircraft was in a region common to the 0.6° radar beam. It means that radar and aircraft were measuring turbulence in approximately the same region at nearly the same time.

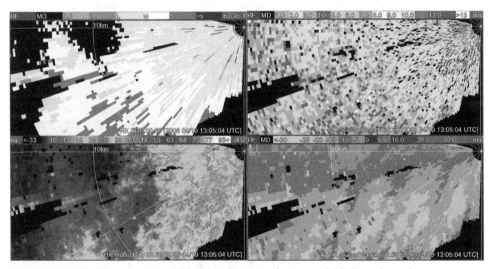

Fig. 12. (a) EDR, (b) spectrum width, (c) reflectivity factor, and (d) Doppler velocity at elevation angle of 0.6° at 13:05 UTC on 19 April 2008. Range ring is at 10 km.

The peaks of these 3 EDR profiles at 13:01, 13:05, and 13:09 UTC are in the green shaded interval and the maximum value is 0.48 $m^{2/3}/s$, just slightly smaller than 0.5 $m^{2/3}/s$. In order to find if there are higher EDR near the flight time (13:05 UTC), we examined the EDR for the two scans one minute before and after the passage of the aircraft at 13:05 UTC in the same volume scan at 13:05 UTC. The profiles are shown with light and dark purple dots in Figure 13. High EDRs with values of 0.69 and 0.76 $m^{2/3}/s$ are found within the shaded interval. This convinces us that the EDR peak is not caused by random error of radar measurements.

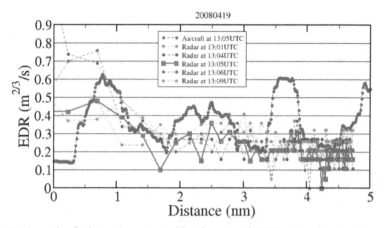

Fig. 13. EDR along the flight path estimated by the aircraft B777 (blue dots) at 13:05 UTC and by the TDWR radar at the time indicated in the legend on 19 April 2008. X axis is the distance between aircraft and the end of runway. The distance interval shaded by the green color indicates where the aircraft passes through the altitude interval observed with the 0.6° elevated beam.

It raises another question: the aircraft may contaminate the radar measurements of the atmospheric status, since the aircraft disturbs the atmosphere and changes the original atmospheric condition in the measurement region as it flies by. In addition, aircraft itself as a target embedded in other scatterers, such as raindrops, may contaminate the spectrum width measurements as well. Both of the two factors could affect spectrum width and EDR value.

It could also be seen that the radar EDR profiles do not match the two aircraft estimated EDR peaks at the distance of 3.65 and 4.90 nm. It might be caused by the spatial difference between the aircraft and the radar beams. The flight heights at the distance of 3.65 and 4.90 nm are higher than the radar beams by about 260 m and 400 m respectively.

Wind shear contribution to spectrum width measurement for this case has been examined. After removing wind shear contribution, the EDR peak at the distance of 0.69 nm is reduced from 0.48 to 0.46 $m^{2/3}/s$ (not shown) at 13:05 UTC. It means that wind shear contribution is small in this region. Because wind shear of the large scale mean wind should be persistent over the 4 minute for entire volume scan, the EDR peaks without wind shear contribution at 13:04 and 13:06 UTC at the distance of 0.69 nm are reduced to 0.67 and 0.74 $m^{2/3}/s$ respectively. It indicates severe turbulence that is matched with aircraft estimate at 13:05 UTC.

Note that the aircraft estimated EDR is considered as ground truth in the above analysis, but it also contains errors and requires significant QC effort, especially as airplane is climbing or descending (Gilbert et al., 2004).

5. Conclusion

This chapter discusses the aviation applications of TDWR. This radar issues microburst alerts which are crucial in the assurance of aviation safety. A typical case of microburst detection by TDWR in association with intense thunderstorms is described first in this chapter. Then the applications of TDWR in the alerting of windshear and turbulence are described. Windshear is alerted through the calculation of windshear hazard factor, which is a rather well established technology. On the other hand, the use of spectrum width data from the radar in the alerting of turbulence has a relatively shorter development history, and the technology is under exploration in Hong Kong.

Study is underway in Hong Kong to use X-band radar in the alerting of windshear and turbulence on experimental basis at the Hong Kong International Airport. The use of long-range S band radar in the alerting of turbulence for enroute aircraft is also under study. Such progress of these studies would be reported in the future.

6. References

Brewster, K.A. and D.S. Zrnic, 1986: Comparison of eddy dissipation rate from spatial spectra of Doppler velocities and Doppler spectrum widths. J. Atmos. Oceanic Technol., 3, 440-452.

Chan, P.W., 2009: Atmospheric turbulence in complex terrain: verifying numerical model results with observations by remote-sensing instruments. Meteorology and Atmospheric Physics, 103, 145–157.

Clark, T.L., T. Keller, J. Coen, P. Neilley, H. Hsu, and W.D. Hall, 1997: Terrain-induced turbulence over Lantau Island: 7 June 1994 Tropical Storm Russ case study. J. Atmos. Sci., 54, 1795-1814.

Cornman, L. B., G. Meymaris, and M. Limber, 2004: An update on the FAA Aviation Weather Research Program's in situ turbulence measurement and reporting system. Preprints, Eleventh Conf. on Aviation, Range, and Aerospace Meteorology, Hyannis, MA, Amer. Meteor. Soc., P4.3.

Doviak, R. J., and D. S. Zrnic, 2006: Doppler radar and weather observations. Dover Publications Inc., Mineola, New York, 562 pp. (except for the preface with links to online errata and supplements, this is an exact copy of the 1st and 2nd printing of the 1993 Academic Press edition).

Fang, M., R.J. Doviak, and Melnikov, 2004: Spectrum width measured by the WSR-88D radar: Error sources and statistics of various weather phenomena. J. Atmos. Oceanic Technol., 21, 888-904.

Gilbert, D., L.B. Cornman, A.R. Rodi, R.G. Frechlich, and R.K. Goodrich, 2004: Calculating EDR from aircraft wind data during flight in and out of Juneau AK: Techniques and challenges associated with non-straight and level flight patterns. Preprints, 11th Conf. on Aviation, Range and Aerospace Meteorology. Hyannis, MA, Amer. Meteor. Soc., CD-ROM, 4.4.

Haverdings, H., 2000: Updated specification of the WINDGRAD algorithm, NLR TR-2000-63, National Aerospace Laboratory, 2000.

Hinton, D.A., 1993: Airborne derivation of microburst alerts from ground-based Terminal Doppler Weather Radar information – a flight evaluation, NASA Technical Memorandum 108990, NASA.

Labitt, M., 1981: Coordinated radar and aircraft observations of turbulence. Project Rep. ATC 108, MIT, Lincoln Lab, 39 pp.

Lee O.S.M. 2004. 'Enhancement of the Anemometer-based System for Windshear Detection at the Hong Kong International Airport.' Eighth Meeting of the Communications/Navigation/Surveillance and Meteorology Sub-Group (CNS/MET/SG/8) of APANPIRG, Bangkok, Thailand, 12 - 16 July 2004. International Civil Aviation Organization.

Merritt, M.W., 1987: Automated detection of microburst windshear for Terminal Doppler Weather Radar, presented at SPIE Conference on Digital Image Processing and Visual Communications Technologies in Meteorology, 27-28 October 1987, Cambridge, MA, USA.

Proctor, F.H., D.A. Hinton and R.L. Bowles, 2000: A windshear hazard index, presented at 9th Conference on Aviation, Range, and Aerospace Meteorology, 11-15 September 2000, Orlando, FL., USA.

Shun C.M. and Chan P.W. 2008. Applications of an infrared Doppler Lidar in detection of wind shear. Journal of Atmospheric and Oceanic Technology 25: 637-655.

Part 2

Precipitation Estimation and Nowcasting

4

Nowcasting

Clive Pierce[1], Alan Seed[2], Sue Ballard[3], David Simonin[3] and Zhihong Li[3]

[1]*Hydro-Meteorological Research, Met Office, FitzRoy Road, Exeter,*
[2]*Centre for Australian Weather and Climate Research, Bureau of Meteorology, Melbourne,*
[3]*Joint Centre for Mesoscale Meteorology, Met Office, Meteorology Building,*
University of Reading, Earley Gate, Reading,
[1,3]*UK*
[2]*Australia*

1. Introduction

The somewhat inelegant term, *nowcasting*, was devised in the mid-1970s (Browning, 1980). It encapsulates a broad spectrum of observation intensive techniques developed for predicting the weather up to a few hours ahead. These techniques are reliant on the rapid processing of high resolution data sets collected by weather radars and satellites. As such, the evolution of nowcasting as a branch of operational meteorology has been closely bound up with post-second world war advances in remote sensing, telecommunications and digital computing. A comprehensive treatment of the subject matter is beyond the scope of this Chapter. In a book about Doppler radar the authors make no apology for focusing on radar based nowcasts of precipitation.

We begin with a brief justification for the use of nowcasts in operational meteorology. This is followed by an overview of nowcasting techniques. A description of some of the key, historical developments in nowcasting is followed by sections on deterministic extrapolation-based nowcasting techniques, errors in precipitation nowcasts and their treatment within nowcasting system frameworks. The remaining sections consider advances in high resolution Numerical Weather Prediction (NWP) model-based nowcasting and review some of the issues and developments surrounding the application of quantitative precipitation nowcasts (QPN) to hydrological forecasting and warning. The Chapter closes with a brief consideration of future prospects for nowcasting.

2. An overview of nowcasting techniques

Operational weather forecasts are produced by primitive equation models known collectively, as Numerical Weather Prediction models. The predictive skill of these models is limited by a number of factors including the accuracy and coverage of routinely available weather observations, the extent to which their model formulations and grid lengths allow the relevant physical and dynamical processes to be modelled accurately, and the non-linear response of the atmospheric system to small perturbations in its state.

Whilst current, operational NWP models are now beginning to resolve important processes such as convection (Lean et al., 2008), their predictive skill generally remains very limited at the convective scales. Furthermore, current computational constraints restrict their operational forecast update cycles to hours, whereas convective phenomena typically exhibit life times of tens of minutes. Thus, NWP-based forecasts of local weather (Browning, 1980) have tended to be rather poor and their use for local forecasting has, until very recently, often been limited to general guidance at the regional scale.

From the 1960s onwards, the availability in near real time of increasingly sophisticated, spatially contiguous, radar and satellite observations, particularly of precipitation or proxies for it, offered the prospect of very short range, local forecasting by extrapolation – the concept of exploiting persistence, either in an Eulerian or Lagrangian reference frame (Germann & Zawadzki, 2002), to make weather predictions with sufficient rapidity to circumvent the perishability of the data. Browning (1980) clarified the relative merits of extrapolation nowcasts and NWP forecasts (see Figure 1), suggesting that the former were of superior accuracy up to 6 hours ahead.

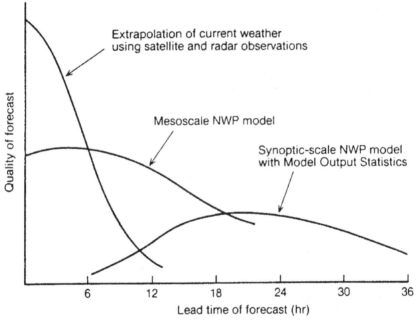

Fig. 1. A schematic diagram after Browning (1980) conceptualizing the relationship between forecasting methodology, skill and forecast range.

The predictability of extrapolation-based precipitation nowcasts and the forecast range at which these nowcasts must hand over to NWP to achieve optimal predictive skill have been explored by a number of authors (e.g. Browning, 1980; Zawadzki et al., 1994; Germann & Zawadzki, 2002; German et al., 2006; Bowler et al., 2006). Recent implementations of convective scale NWP model forecasts are now reducing the useful range of extrapolation-based nowcasts to a few hours ahead, as discussed later in this Chapter.

In the following section, we describe some of the key milestones in radar-based precipitation nowcasting and review these in the context of parallel advances in relevant areas of science and technology.

3. Radar-based nowcasting – A brief history

3.1 Origins of weather radar, and early research

Operational weather radar has its origins in the development of military radar during World War Two. The invention of the resonant cavity magnetron by John Randall and Harry Boot at the University of Birmingham in England in 1940 allowed the construction of high powered, centimeter-band radars, suitable for detecting precipitation. The sharing of this technology with American scientists early in the 1940s facilitated its subsequent development for meteorological applications.

Important early papers include those on rain drop size distributions (Marshall & Palmer, 1948) and shapes (Browne & Robinson, 1952; Hunter, 1954; Newell et al., 1955), the measurement of precipitation (Ryde, 1946; Byers, 1948; Bowen, 1951; Twomey, 1953; Battan, 1953; Stout & Neill, 1953), its vertical structure (Langille & Gunn, 1948) and associated estimation errors (Hitschfeld & Bordan, 1954), and those on thunderstorm identification, behaviour and dynamics (Wexler & Swingle, 1947; Byers & Braham, 1949; Wexler, 1951; Ligda, 1951; Battan, 1953).

3.2 Extrapolation techniques

The concept of extrapolating radar echoes for the short term prediction of precipitation was first proposed by Ligda (1953). The earliest demonstration of the application of objective extrapolation to radar echoes is described by Hilst and Russo (1960), whilst Noel and Fleischer (1960) were amongst the first radar meteorologists to explore the predictability of precipitation echoes using this approach. Further noteworthy papers are those published by Russo and Bowne (1962) and Kessler and Russo (1963). Kessler (1966) and Wilson (1966) explored the use of cross correlation statistics to diagnose a best estimate of echo pattern average motion. Wilson (1966) used the maximum value of the cross correlation coefficient as an indicator of pattern development.

Two important conclusions were drawn from these early studies. The first of these was the positive correlation between the predictability of precipitation features and their size: large features tend to be longer lived than small ones. The second conclusion is an adjunct to the first, namely that small scale features are generally short lived – typically a few tens of minutes. These findings are consistent with early investigations into the multi-scaling properties of the atmosphere and associated limits on atmospheric predictability (Lorenz, 1963; 1973).

The 1970s saw the further development of cross correlation-based nowcasting algorithms and their automation. Zawadzki (1973) developed an optical device for measuring the space-time statistical properties of radar inferred precipitation fields. Austin and Bellon (1974) evaluated an automated, computerized pattern matching programme for nowcasting precipitation up to 3 hours ahead. They concluded that the useful range of these nowcasts varied with the nature and extent of the precipitation. Nonetheless, this approach was shown to be consistently skilful up to one hour ahead over a wide range of events.

This latter work led to the operational implementation of an algorithm based upon global cross correlation at McGill University in the mid-1970s. Bellon and Austin (1978) reviewed the operational performance of this scheme, known as SHARP (Short-Term Automated Radar Prediction), on two years' worth of data. The experience gained allowed subsequent enhancement of their cross correlation method to enable independent tracking of different echoes (Austin & Bellon, 1982) using a nine vector motion field. Rinehart (1981) describes a similar, multi-vector, cross correlation approach to determine and extrapolate the motion of individual storms within a multi-storm system.

3.3 Cell tracking

Algorithms founded on the tracking of radar echo centroids evolved in parallel with field-based pattern matching techniques. These were developed specifically for nowcasting thunderstorms, initially in North America. Amongst the earliest of these echo centroid trackers were those described by Wilk and Gray (1970) and Zittel (1976). The extrapolation vectors were diagnosed using a linear least squares fit through successive positions of the echo centroids. Duda and Blackmer (1972) and Blackmer et al. (1973) formulated clustering techniques to resolve difficulties in cases involving the merging and splitting of echoes.

Refinements to these early techniques were subsequently developed and implemented within operational tools during the following decades. Several good examples are the Storm Cell Identification and Tracking (SCIT) algorithm (Witt & Johnson, 1993) and the Thunderstorm Identification, Tracking, Analysis and Nowcasting (TITAN) system (Dixon & Wiener, 1993).

3.4 Steady state versus growth and decay

The proto-type, operational nowcasting algorithms developed during the 1970s were generally reliant on the steady state assumption. Tsonis and Austin (1981) explored echo size and intensity trending with a view to improving the prediction of long lived convective cells. They found negligible improvement in skill, even using sophisticated non-linear time trending schemes. Wilson et al. (1998) drew similar conclusions in a study involving the use of the TITAN system (Dixon and Wiener, 1993). These results are consistent with the findings of theoretical and NWP modelling experiments showing that the evolution of convective scale features in the atmosphere is non-linear and, to a degree, chaotic (Tsonis, 1989).

3.5 Fractal properties of precipitation

During the 1980s and 1990s, an improved understanding of the chaotic influence of atmospheric processes such as turbulence on the predictability of precipitation was reflected in a growing number of publications exploring the so called *scaling* or *multi-fractal* attributes of meteorological fields, including those of radar derived precipitation fields. Scaling behaviour or self-similarity implies that similar features can be observed in the atmosphere over a wide range of space and time scales, and that the relationship between certain statistical attributes of a precipitation field measured at different scales can be described by equations which incorporate a scaling factor. The formative papers in this area include those by Lovejoy and Schertzer (1985, 1986).

3.6 Exploration of statistical and statistical-dynamical models of precipitation

This same period saw the development of a number of statistical (Krajewski & Georgakakos, 1985; Cox & Isham, 1988) and statistical-dynamical predictive models of precipitation (Georgakakos & Bras, 1984a, b; Lee & Georgakakos, 1990; French & Krajewski, 1994; Bell & Moore, 2000a). The latter were formulated to assimilate radar estimates and surface observations of precipitation and make forecasts using a simplified treatment of the governing atmospheric equations, focusing on the conservation of water mass. These studies were motivated by the operational forecasting requirements of the hydrological community and the limitations of "steady-state" nowcasting techniques and the first generation of operational, mesoscale NWP models.

3.7 Impact of forecast uncertainties

A growing recognition of the need to account for and communicate meteorological forecast uncertainty (Murphy & Carter, 1980; Krzysztofowicz, 1983), particularly in relation to precipitation, led to the development of a range techniques for probabilistic precipitation nowcasting. Andersson and Ivarsson (1991) evaluated an advection-based nowcasting scheme in which the probability of precipitation at a given location is estimated from the areal distribution of precipitation in a neighbourhood surrounding it (see also Schmid et al., 2000) This approach accounts for the impact of extrapolation errors on the location of advected precipitation. Other authors have adopted similar approaches. For example, Germann and Zawadzki (2004) used a local Lagrangian method to produce probabilistic extrapolation nowcasts.

The previously mentioned theoretical work on multi-fractals, and empirical studies supporting a scaling model representation of precipitation fields, laid the foundations for the development of a number of stochastic precipitation nowcasting schemes exploiting scale decomposition frameworks. Seed (2003) adopted a multi-scale decomposition framework in his S-PROG (Spectral-Prognosis) scheme to nowcast the space-time evolution of high resolution radar derived precipitation fields (see Figure 2); he highlighted the potential application of S-PROG to conditional simulation and design storm modelling.

In a similar vein, the McGill Algorithm for Precipitation Nowcasting by Lagrangian Extrapolation (MAPLE; Turner et al., 2004) exploits a wavelet transform to model the predictability of precipitation as a function of scale. The aim of the scale decomposition is to filter out the unpredictable scales in an extrapolation nowcast, and in so doing, minimize the Root Mean Square nowcast error (typically measured using rain gauge observations and/or radar inferred estimates of surface precipitation rate or accumulation).

Pegram and Clothier (2001) used a power law model to filter Gaussian distributed random numbers to generate stochastic realizations of radar precipitation fields in their String of Beads Model (SBM). Noise generation techniques similar to these were combined with a stochastic model of extrapolation velocity errors in the Short Term Ensemble Prediction System (STEPS, Bowler et al., 2006) to produce operational precipitation nowcasts quantifying uncertainties in phase as well as amplitude. In STEPS, the noise serves several purposes: it enables ensembles of equally likely nowcast solutions to be generated by perturbing predicted features as they lose skill; it also downscales an NWP forecast, injecting variance at scales lacking power (variance) relative to the radar.

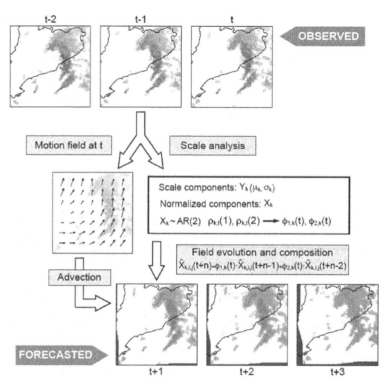

Fig. 2. The generation of an extrapolation nowcast using the Spectral-Prognosis (Seed, 2003) multi-scale decomposition (cascade) framework (after Berenguer et al., 2005) . The motion of the precipitation field is derived from radar inferred analyses of precipitation rate valid at t and t-1 (typically a 10 or 15 minute time step). The temporal evolution of the extrapolated field is modelled on the hierarchy of scales produced by the cascade decomposition, using a hierarchy of second order auto-regressive (AR-2) models – one for each scale – and the analyses of precipitation rate valid at t, t-1 and t-2.

3.8 Improvements in extrapolation techniques

The past two decades have also seen further refinements to the extrapolation schemes exploited by precipitation nowcasting algorithms, notably in the form of COTREC (Li et al., 1995; Mecklenburg et al., 2000), Variational Echo Tracking (VET; Germann & Zawadzki, 2002) and optical flow (Bowler et al., 2004, Peura & Hohti, 2004). COTREC constrains the cross correlation diagnosed displacement vectors using the two dimensional continuity equation. This is equivalent to minimizing the divergence of velocities derived for adjacent blocks. The benefits over TREC (Rinehart and Garvey, 1978) were shown to be due to the elimination of spurious motion vectors caused by clutter, beam blockages and rapid changes in the precipitation pattern. In common with optical flow, the VET scheme diagnoses a field of motion by direct application of the optical flow constraint equation. Bowler et al. (2004) solve this equation before applying a smoothness constraint where as Germann and Zawadzki (2002) uses a conjugate gradient method to minimize residuals from two constraints simultaneously.

3.9 NWP-based nowcasting

The 1990s saw the first attempts to run convection resolving NWP model forecasts assimilating radar data (Lin et al., 1993). These early experiments were focused on predicting convective storms.

Some success was demonstrated in cases involving convection strongly forced by the large scale environment. However, other studies showed that convective initiation in a weakly forced environment is difficult to predict because the location and timing of initiation are very sensitive to variations in low level temperature and moisture. It seems likely that the ability of convection resolving NWP models to predict convection is a function of the predominant scale of the associated forcing. In the UK, much of the convection is forced by small-scale orography, as demonstrated by the Convective Storms Initiation Project (Morcrette et al., 2007; Lean et al., 2008).

Despite these challenges, Lean et al. (2008) found that convection resolving models performed better in terms of convective initiation than a 12 km grid length model with parameterized convection, and a number of national weather services are now running operational, convection resolving NWP models. Indeed, some are trialling configurations with hourly or sub-hourly assimilation of radar data. NWP nowcast experiments in the UK show some improvements in NWP forecast skill in the nowcast time frame. Prospects for NWP nowcasting will be discussed in more detail later in this Chapter.

4. Conventional nowcasting techniques

4.1 Deterministic techniques

4.1.1 Cell tracking

Cell trackers or object-based nowcasting schemes are typically developed in areas where severe convective storms are a significant hazard, and are best suited to the generation of qualitative warnings of severe convective weather. In general, object-based algorithms are used to predict the location of a (convective) object in the future and thereby assign the properties of the object to that location. For example, a storm might be deemed to contain large hail, and therefore a warning of large hail will be issued for the locations on the forecast storm track.

The basic elements of cell tracking are:

1. devise a set of rules that will be used to identify the bounds of an object in either two or three dimensions;
2. analyse current data to identify objects and assign attributes to them (heavy rain, damaging wind, large hail etc);
3. link the objects to existing tracks and estimate the advection velocity;
4. predict the location of objects in the future.

Most cell tracking algorithms define an object, either as a set of contiguous points that exceed some threshold in radar reflectivity, typically 35, 40 or 45 dBz (e.g. Dixon & Weiner, 1993; Han et al., 2009), or as a small region of increased reflectivity (Crane, 1979), or both (e.g. Handwerker, 2002). Defining the object in three dimensions allows one to compute the volume and height of the cell. This adds value when assigning the elements of severe weather or some sort of severity index, but does not necessarily add value to the identification and tracking of the cells. Rigo et al. (2010) used both 2D and 3D radar products and total lightning data to identify and track convective storms. A storm track is defined as a time series of cell positions. Assigning cells to tracks is the most complex aspect of these algorithms.

All cell tracking algorithms have to deal with cell initiation, mergers, splits, and terminations – the hatches, matches, and dispatches as it were – and this is often the point of differentiation between the various approaches. Errors in assigning the correct cell to a track are a major cause of error when estimating the cell velocity. The size of the object depends on the threshold that has been selected. Therefore, the predictability of the object decreases as the threshold is increased since the lifetime of a cell is related to its size. Using a high threshold to define the cell will make it more difficult to assign a cell to a track. This will increase the errors when estimating the track velocity. Using a low threshold will increase the longevity of the tracks, but will tend to limit the ability to forecast the location of the most severe cells within the storm. The concept of an "object" becomes less useful as the precipitation becomes more widespread. At some point (depending on the skill of the tracking algorithm), cell tracking algorithms fail to provide useful forecasts.

TITAN (Dixon & Weiner, 1993) and SCIT (Johnson et al., 1998) are good examples of what can be achieved in the object-tracking paradigm. Both TITAN and SCIT use the three-dimensional radar reflectivity data to identify a convective object that is defined by a reflectivity threshold. In TITAN, the current objects are linked to past objects through combinatorial optimization. This minimizes the total advection and change in cell volume between the previous and current time steps. Many other cell tracking algorithms, SCIT for example, assign the cell that is closest to the forecast location of an active track. Han et al. (2009) evaluated several extensions to TITAN including improvements in assigning cells to tracks and using TREC motion vectors to advect the cells. In assigning a cell to a track, they found that the most significant improvements were due to adding a requirement that the forecast cell from the track at the previous time step must overlap with the current cell.

4.1.2 Field-based advection

Field tracking algorithms generally divide a Cartesian grid of radar reflectivity or rain rate into a number of tiles and then find the advection of the tile that maximizes the cross correlation (or some other measure of similarity) between successive time steps in the data. The mean advection vector for each tile containing rain is then calculated by applying some form of constraint to minimize the divergence of the resulting vectors.

A number of the current field tracking-based nowcasting algorithms use COTREC (Li et al., 1995) as the basis for deriving the advection vectors. Examples include the system that has been developed at the Czech Hydrometeorological Institute (Novak, 2007), the Hong Kong Observatory system, SWIRLS (Li et al., 2000), and the system implemented at the Guangdong Meteorological Observatory system (Liang et al., 2010). Liang et al. (2010) determined that the optimum size of the tile was 30 km. Li et al. (2000) evaluated the performance of an advection scheme on a 93 x 93 grid using a 19 pixel tile: this equates to 20 km on their 256 km x 256 km domain.

Bowler et al. (2004) used the optical flow constraint (Horn & Schunck, 1981) approach that is used for computer vision applications to derive the mean advection vector for tiles with rain. Optical flow uses least squares to find the (u,v) that minimizes the two-dimensional conservation equation

$$\frac{dR}{dt} = u\frac{\partial R}{\partial x} + v\frac{\partial R}{\partial y} + \frac{\partial R}{\partial t} = 0 \qquad (1)$$

over a local neighbourhood.

Bowler et al. (2004) smoothed the field using a moving average over a (15 x 15) pixel mask before calculating the partial derivatives using a finite difference scheme. The smoothed image was then partitioned into 48 x 48 km² tiles and least squares used to estimate the mean advection vector within the tile. The resulting vectors were then smoothed so as to minimize $\nabla^2 V$. Grecu and Krajewski (2000) used a similar approach over 40 x 40 km² tiles. Foresti and Pozdnoukhov (2011) used optical flow to track areas with rain rates that exceeded 10 mm/h. Essentially, this represents the application of optical flow to cell tracking.

Germann and Zawadzki (2002) used the Variational Echo Tracking (VET) method of Laroche and Zawadzki (1995) to derive the advection velocities. This technique partitions the field into small tiles and then uses the conjugate gradient method to minimize a cost function in one global minimization. The cost function includes a smoothness term. The difference between this approach and optical flow is that optical flow applies the smoothness constraint after the velocity field has been calculated for each tile, thereby avoiding an expensive global minimization (Bowler et al. 2004). Ruzanski et al. (2011) describe another approach using a linear least squares technique in the frequency domain.

Cell tracking algorithms assign a velocity to each object and this is advected with a constant velocity during the forecast period. Such an approach is not optimal for field tracking algorithms because it does not allow for changes in direction and speed of motion during the forecast period. Germann and Zawadzki (2002) undertook a detailed analysis of several advection algorithms and found that a modified semi-Lagrangian backward interpolation scheme was optimal. Bowler et al. (2004, 2006) used the simpler semi-Lagrangian scheme that is applied for each time step in the forecast time series. Semi-Lagrangian advection requires a velocity at each pixel in the field and the optical flow technique does not provide advection vectors for tiles that have no rainfall. Therefore the velocity at each pixel must either be interpolated from the tiles with rain, or provided by a hierarchical approach that progressively reduces the size of the tiles that are used in the analysis (e.g. Germann and Zawadzki, 2002).

Kernel-based methods have been employed for advection by Ruzanski et al. (2011) and Fox and Wikle (2005) using

$$\mathbf{y}_{t+1} = \mathbf{H}\mathbf{y}_t \qquad (2)$$

where

$$\mathbf{y}_t = [y(s_1, t), y(s_2, t), , , y(s_n, t)]^T \qquad (3)$$

is the vector of the n pixels in the image and $\mathbf{H} = \{h_{ij}\}$ is the $n \times n$ matrix of the advection operator.

Ruzanski et al. (2011) report that their approach is computationally efficient, although the time taken to derive the motion vectors was comparable to that required for optical flow. Furthermore, the advection algorithm was an order of magnitude slower than a simple implementation of a semi-Lagrangian backward interpolation scheme. Both Ruzanski et al. (2011) and Fox and Wikle (2005) demonstrated their methods using small images. The size of the advection operator is likely to become a constraint when using this technique to advect a large (say 10^6 pixels) image.

4.1.3 Analogues

Panziera et al. (2011) provide a good introduction on the assumptions and use of analogues in nowcasting. Advection-based tracking techniques rely on the assumption that precipitation fields evolve relatively slowly in Lagrangian coordinates and, therefore, their future state can be predicted largely by extrapolation. The assumption of Lagrangian persistence becomes a major limitation on the accuracy of nowcasts in situations where a field evolves rapidly, for example in situations where new storms are initiated or existing storms grow or decay. Data mining and analogue techniques seek to predict initiation, growth and decay by matching the current weather pattern with similar, past events and then use these past events as the basis for generating a forecast.

The first step in the use of analogues is to, either identify a set of regimes in the historical data, or identify a set of predictors that can be used as measures of similarity. Thereafter, the analogue that is closest to the current situation is selected and used as a basis for the forecast. This implies that the technique must be trained for each location, and that a significant historical record is available. Panziera et al. (2011) used predictors of mesoscale airflow and air-mass stability to select 120 analogues, and then employed two measures from the radar derived rainfall fields to select a set of 12 analogues to use as a forecast ensemble. Foresti and Pozdnoukhov (2011) derived maps of where orographic enhancement was likely to occur for a set of weather types. These could then be used to correct biases in advection forecasts.

4.2 Errors in precipitation nowcasts

4.2.1 Error sources and attribution

Sources of forecast errors include errors in the initial quantitative precipitation estimates (QPE), those arising from incorrect diagnosis of the field of motion, and changes in the motion and evolution of precipitation fields during the forecast period.

Approximately half of the total forecast error in the first hour of a forecast is due to errors in the radar derived rainfall analyses (Bellon & Austin, 1984; Fabry & Seed, 2009). This is because radar rainfall estimation errors, arising from variations in the relationship employed to convert the observed radar reflectivity to rainfall, have significant correlations over about an hour in time (Lee et al., 2007) and tens of kilometres in space (Velasco-Forero et al., 2009; Yeung et al., 2011).

Dance et al. (2010) used a year of TITAN tracks to investigate how cell tracking errors varied as a function of lead time, storm intensity, speed and duration. They found that the RMS errors in track speed and direction over the year were about 10 km/h and 30° respectively. Dance et al. (2010) found that tracking errors (both speed and direction) were large when the track speeds were less than 15 km/h; also, errors in track direction decreased with increasing speed.

Mecklenburg et al. (2000) investigated the tracking errors for TREC and COTREC and found that the mean absolute displacement and direction error for a 30-min forecast of convection was about 10 km and 20° respectively. Ebert et al. (2004) showed that TITAN cell tracking algorithms had a median error of about 10 km/h for intense cells. The median tracking error for the baseline field tracking algorithm – a correlation technique finding a single advection vector for the entire field of convective storms – was found to be 20 km/h using the same data.

Hourly accumulations of rainfall typically have correlation lengths of the order of 10 km (e.g. Anagnostou et al., 1999; Gebremichael & Krajewski, 2004). The tracking error after an hour is at least the same order of magnitude as the correlation length of the accumulations. Therefore, one would expect that enhancements to the current tracking algorithms should lead to improvements in the skill of nowcasts at lead times when tracking errors become a significant fraction of the correlation length of the rainfall field: this is likely to be around T + 30 minutes.

Berenguer et al. (2005) found that the temporal evolution of the advection field was not a significant source of error for nowcasts with lead times less than 60 minutes. Bowler et al. (2006) discovered that forecast errors due to the temporal evolution of the advection field were negligible in the first three hours of a nowcast and accounted for 10% of the total error after six hours.

It is interesting to note that the probability distribution of cell tracking errors is highly skewed (see, for example, Figure 20 of Ebert et al., 2004) and that the maximum 60 minute location error can be as high as 70 km. The fat tail in the distribution of tracking errors is a significant issue for operational nowcasting systems. Manual editing of the tracks (e.g. Bally, 2004) adds value to the automatic forecasts by eliminating the tracks that are regarded by the forecasters as being, either unimportant from a severe weather perspective, or incorrect.

Errors due to the initiation and decay of storms during the forecast period become increasingly dominant as the lead time extends beyond 60 minutes (Wilson et al., 2010). Zawadzki et al. (1994) evaluated the limits of predictability of rainfall fields as a function of space and time and found that the time for a Lagrangian persistence forecast to reach a correlation of 0.5 ranged from 40 to 112 minutes. They also found that these predictability times depended on the scales present in the rainfall field.

4.2.2 Space-time structure of errors and their treatment

Roca-Sancho et al. (2009) examined the spatial and temporal structure of forecast errors for MAPLE. They demonstrated that the temporal correlation of forecasts errors was very low after 60 minutes and that the spatial structure of the forecast errors progressively resembled that of rainfall with increasing lead time. The latter effect was due to increasing errors in the location of rainfall. Fabry and Seed (2009) showed that forecasts of high rain rates were generally over-predictions and that the performance of advection forecasts in the recent past is not a good predictor of future performance. The best predictors were found to be raining fraction and the rate of change in mean areal precipitation over the forecast domain.

Germann and Zawadzki (2002) demonstrated that filtering the rainfall analysis field with a 64 km, low-pass filter increased Lagrangian life times by between 40 and 60 minutes, depending on the extent to which small scale features are embedded in larger-scale rain areas. Germann et al. (2006) state that the upper bound for an advection-based nowcasting system that does not include growth and dissipation of rainfall is about six hours. The typical lifetime of a storm is closely related to the scale of the storm, and is often represented as a power law of the scale (e.g. Marsan et al., 1996; Schertzer et al., 1997; Seed et al., 1999). Therefore, some nowcasting systems improve the accuracy (in the RMS error sense) of their predictions by progressively smoothing out the small scale features present in the analysis field (e.g. Seed, 2003; Turner et al., 2004). This removes features from the nowcast that are essentially unpredictable.

An alternative way of handling the perishability of the fine scale components in advected precipitation fields is to model them stochastically. This approach will be discussed in section 4.3.

4.2.3 Performance of nowcasting algorithms

The Critical Success Index is often used to report the accuracy of nowcasting algorithms presented in the literature. Ruzanski et al. (2011) found a CSI of approximately 0.5 after 10 minutes at a spatial resolution of 0.5 km. Liang et al. (2010) calculated a CSI of approximately 0.35 after 60 minutes for echoes in the 15-45 dBZ range at 2 km resolution. Berenguer et al. (2011) report a CSI for 60 minute forecasts of reflectivity (dBZ) at 1 km resolution of approximately 0.5 for widespread rainfall, and in the range of 0.1 to 0.3 for isolated convection. Poli et al. (2008) discovered that the CSI was generally low at the start and end of a storm, reaching a peak of around 0.4 for 1 km resolution T+60 minute forecasts of reflectivity greater than 30 dBZ.

Nine nowcasting systems were implemented for the Sydney 2000 Forecast Demonstration Project (Ebert et al., 2004) and eight nowcasting systems participated in the Beijing 2008 Olympics' Forecast Demonstration Project (Wang et al., 2009; Wilson et al., 2010). Wang et al. (2009) demonstrated that the overall performance of the nowcasting systems had improved during the years from 2000 to 2008. They showed that the maximum CSI for forecasts of hourly precipitation accumulation greater than 1 mm/h increased from 0.2 in 2000 to 0.45 in 2008, although the maximum CSI for rain greater than 10 mm/h was still only 0.15.

Lee et al. (2009) found that the CSI decreased with increasing rain rate and forecast lead time: the CSI for 60 minute rainfall forecasts decreased from 0.60 for 0.1 mm/h to 0.2 for 10 mm/h rain rates. Ebert et al. (2004) reported that the CSI for rain greater than 20 mm/h is essentially zero. This implies that the use of nowcasting techniques to predict the precise location of extreme rain for flash flood warning may not be viable.

In summary then, the accuracy of a nowcast depends on the accuracy of the initial radar derived rainfall field, the degree of spatial organization of the rain, the rain rate, and forecast lead time. Also, it is likely to be higher in the middle of the storm (in both space and time) than at the edges.

4.3 Probabilistic techniques

4.3.1 Justification

Given the magnitude of the errors in a 30 minute precipitation nowcast, it is reasonable to adopt a probabilistic approach to nowcasting and attempt to convey to the users the uncertainty that is associated with a particular weather situation. As explained earlier, within the extrapolation nowcast framework, errors can be categorized into those attributable to the radar observations and processing, inaccuracies in the field of motion used to advect the observations, and errors arising from assumptions made about the Lagrangian evolution of the advected precipitation field.

4.3.2 Methods of handling uncertainties

A number of techniques have been developed for modelling nowcast errors with a view to producing probabilistic precipitation nowcast products. One of the simplest entails time-lagging a consecutive series of deterministic nowcasts using techniques similar to those demonstrated in a NWP post-processing context (Mittermaier, 2007). Each member of the time-lagged ensemble is assigned a weight which is a function of lead time. SWIRLS generates probabilistic nowcasts using this approach (Wang et al., 2009).

Another approach relies on the assumption that errors in the diagnosed advection velocity field predominate. Consequently, the probability of exceeding a chosen precipitation threshold at a given location can be derived from the distribution of precipitation in a neighbourhood surrounding the forecast location. The neighbourhood size increases with lead time to reflect to the growth in advection errors (Andersson & Ivarsson, 1991; Schmid et al., 2000).

Germann and Zawadzki (2004) compared four methods of generating probabilistic precipitation nowcasts based upon radar extrapolation. They concluded that the most skilful method was one based upon the local Lagrangian technique. Essentially, this produces an advection forecast using a semi-Lagrangian backward advection scheme and then uses the probability distribution of forecast rain rates in some search area centred on a pixel to calculate the probability of exceeding a threshold at that location. The size of the search area increases with lead time to reflect the increasing forecast uncertainty. This approach has since been exploited by others, for example Megenhardt et al. (2004), and more recently, Kober et al. (2011).

Other authors have focused their attentions on modelling errors using stochastic space-time models. Pegram and Clothier (2001) used a power law model to filter Gaussian distributed random numbers to generate stochastic realizations of radar precipitation fields in their String of Beads model (SBM). Noise generation techniques similar to these were combined with a stochastic model of extrapolation velocity errors in the Short Term Ensemble Prediction System (STEPS; Bowler et al., 2006) to produce operational precipitation nowcasts quantifying uncertainties in phase as well as amplitude. In STEPS, the noise serves several purposes: it enables ensembles of equally likely nowcast solutions to be generated by perturbing predicted features as they lose skill; it also downscales an NWP forecast, injecting variance at scales lacking power (variance) relative to the radar.

4.3.3 Treatment of observation errors

Uncertainties in nowcasts of precipitation also derive from errors in the radar observations and processing. Austin (1987) categorized radar errors into physical biases, measurement biases and random sampling errors. Historically, much effort has been invested in improving deterministic estimates of precipitation accumulation at the surface by correcting physical (e.g. ground clutter and beam blockage) and measurement (e.g. Z-R conversion) biases. However, more recently, a growing number of researchers have focused their attentions on the treatment of random sampling errors and how these can be utilized within stochastic, integrated system frameworks to improve hydro-meteorological nowcasting.

Two main approaches to the modelling of random sampling errors in QPE have been described in the literature: one entails a statistical description of the difference between the radar estimates and a reference (e.g. Ciach et al., 2007; Llort et al., 2008; Germann et al., 2009); a second involves modelling the characteristics of individual sources of error (e.g. Jordan et al., 2003; Lee & Zawadzki, 2005a, 2005b, 2006; Lee, 2006; Lee et al., 2007). The challenge with the first approach is the need for a reference field: this is usually derived from a dense network of rain gauges. The difficulty with the second approach is that the true error structure of QPEs can vary significantly depending on the meteorological conditions and is therefore largely unknowable.

Germann et al. (2009) describe a radar ensemble generator using LU decomposition (factorization) of the radar-gauge error covariance matrix to derive an ensemble of precipitation fields. Each ensemble member is the sum of the bias corrected, deterministically derived radar precipitation field and a stochastic perturbation representing the random error. The stochastic term is generated such that it preserves the correct space–time error covariances. The authors present the results of the coupling of a real-time implementation of the radar ensemble generator with a semi-distributed hydrological model.

Norman et al. (2010) implemented several radar ensemble generators and compared their performance on a selection of case study events using rain gauges. An implementation of the Germann et al. (2009) scheme was found to be marginally superior to one comprising separate models of Z-R (Lee et al., 2007) and VPR (Jordan et al., 2003) errors. Pierce et al. (2011) integrated these two ensemble generators to produce ensembles of radar-based analyses of surface precipitation rate for input to STEPS. They evaluated the impact of these ensembles on the performance of STEPS ensemble precipitation nowcasts. Verification results demonstrated that accounting for QPE errors improved the ensemble spread-skill relationship in the first hour of the nowcasts.

One alternative to the stochastic QPE and QPN schemes described above is the use of historical analogues. Panziera et al. (2011) describe an analogue-based heuristic tool for nowcasting orographically forced precipitation. The system known as Nowcasting of Orographic Rainfall by means of Analogues, exploits the strong correlation between orographic rainfall and predictors describing mesoscale flow and air mass stability, to identify past events with predictors similar to those derived from real time observations. The authors present verification results showing that NORA performs better than Eulerian persistence for nowcasts with lead times of more than an hour.

5. NWP-based nowcasting

5.1 Introduction

In the past few years, increasing availability of high powered computers and the implementation of non-hydrostatic models have made NWP at the convective scales (1 km–4 km horizontal grid length) a reality for national weather services. Many centres are already running these models operationally with update cycles of between 3 and 6 hours to generate short-range forecasts up to about T+36 hours. Traditionally, these forecasts have been deployed in combination with nowcasting techniques to deliver optimal guidance. However, recently, centres have begun to explore the use of NWP-based systems for nowcasting.

5.2 The challenges

For nowcasting purposes, the key component of NWP is the data assimilation of high resolution observations in space and time, especially radar and geostationary satellite data. Traditional nowcasting techniques use these observations to produce forecasts of rain, cloud and associated weather with observation derived advection velocities, or NWP forecast wind fields, or a combination of both. Nowcasts are also produced from analyses of other weather elements including screen temperature, visibility, 10 m wind and wind gusts. However, these systems do not use the observations in an optimal manner and may not use all available observation types.

Data assimilation into NWP models potentially offers the ability to use all observations in a consistent and synergistic manner to provide the best estimate of the state of the atmosphere from which to produce a nowcast. At this time, nudging, variational data assimilation (3D-Var and 4D-Var) and ensemble Kalman filters (EnKF; Sun, 2005b) for high resolution data assimilation are being used in weather services or are under development in research centres around the world. Indeed, some national weather services are already running operational NWP models with data assimilation at grid lengths in the range 1 km-10 km. Most of this work relies heavily on the exploitation of Doppler radar measured radial winds and reflectivity data or derived surface rain rates.

One challenge for NWP-based nowcasting is to match the skill of traditional methods in the first two hours. Traditional nowcasts closely fit the observations because they employ extrapolation techniques and so use the observations themselves (i.e. radar derived surface rain rate) at analysis time. This is challenging for NWP because unresolved scales are excluded from the model state, data assimilation systems are designed, not to match observations, but to achieve a good and balanced forecast over a longer period of time, and the T+0 fields from the NWP system are essentially a weighted fit to both the NWP forecast and the observations.

Also, traditional nowcasts can produce forecasts within a few minutes of data time, but complex data assimilation methods and numerical integration of the governing atmospheric equations are more costly and therefore take longer. However, if these techniques produce improved forecasts at longer lead times, the benefits outweigh the timeliness issue and reduced accuracy in the first 2 hours.

Another performance issue with NWP-based nowcasts relates to the latency of the boundary conditions. This arises because domain sizes are usually small and are nested in coarser resolution forecasts or larger domain forecasts with less frequent analysis cycling and later data cut-off times. The consequences are that the boundary conditions and synoptic scale forcing cannot be refreshed as frequently or as recently on the larger domain(s) as they are on the nowcast inner domain. This limits the skill at longer forecast ranges and possibly close to the boundaries.

Nonetheless, the advantage of NWP-based nowcasting lies in the fact that model formulation, dynamical equations and physical parameterizations can predict the non-linear evolution of weather elements and, in particular, the generation and decay of precipitating weather systems.

5.3 A status report

To investigate the direct use of NWP for nowcasting, the Met Office in the UK is developing an hourly cycling 4D-Var high resolution (1.5 km) NWP system to run on a domain covering southern England (see section 5.4). This is nested within the most recent forecasts for the whole of the UK (1.5 km resolution forecasts produced every 6 hours from 3 hourly 3D-Var data assimilation cycles at 3 km resolution) to obtain boundary conditions. The latter may be up to 6 hours old. Although 4D-Var is more expensive than 3D-Var, the aim is to evaluate the benefit of assimilating high time-frequency sub-hourly data (Ballard et al., 2011): see section 5.4 for more details.

Over the past 20 years, NCAR has undertaken many studies to explore the assimilation of radar data into high resolution cloud and NWP forecast models. These have included using the Variational Doppler Radar Assimilation System (VDRAS – Sun, 2005a, 2005b; Sun & Crook, 1994, 1997, 1998, 2001; Sun & Zhang, 2008) with 4D-VAR (Sun et al., 1991, 2012). These tend to use very short time-windows and have exploited the mesoscale model, MM5 3D-Var (Xiao et al., 2005) and the Weather Research & Forecasting Model (WRF) 3D-Var and 4D-VAR, or ensemble Kalman filter (Caya et al., 2005). These were run using VDRAS as part of the forecast demonstration project during the Beijing Olympics (Sun et al., 2010).

Meteo-France has a 2.5 km, 3-hourly cycling 3D-Var scheme covering France (the Application of Research to Operations at Mesoscale – AROME-France). This has been operational since December 2008 (Seity et al., 2011; Brousseau et al., 2011). Radial Doppler winds (Montmerle & Faccani, 2009) and humidity profiles derived from radar reflectivity (Caumont et al., 2010) are assimilated. Meteo-France is also undertaking a project entitled, "AROME-Nowcasting", to adapt their 2.5 km grid length model, AROME, to meet the requirements of nowcasting. The main difference to AROME-France is the production of an analysis every hour, but without cycling. The potential benefits of a system called AROME-airport, based at Charles de Gaulle airport near Paris, are also being explored. This model will provide an input to a Wake-Vortex forecast model. The main goal is to add new, dedicated observations, and to run a 500 m grid length model in a configuration comparable with conventional nowcasts (Ludovic Auger, MeteoFrance, personal communication WMO/WWRP Workshop on Use of NWP for Nowcasting, Boulder 2011).

DWD has a 2.8 km forecast model with a nudging assimilation scheme (Consortium for Small-scale Modeling, COSMO) covering Germany (COSMO-DE, Stephan et al., 2008) and is developing PP KENDA (Priority Project "KENDA" – Km-scale Ensemble-based Data Assimilation) for a 1 km-3 km scale Ensemble Prediction System known as LETKF (Local Ensemble Transform Kalman Filter; Ott et al., 2004). MeteoSwiss is running a 2 km version of COSMO. Various collaborating meteorological services are running, or are planning to run versions of these systems.

The HIRLAM (HIgh Resolution Limited Area Model) European community run their 3D-Var system (Gustafsson et al., 2001) with the HIRLAM model at grid lengths down to about 3.3 km and are developing a new HARMONIE system to run at about 2.5 km. They have also run experiments comparing three hourly and hourly cycling at 11 km and are exploring the impact of GPS and Doppler radar radial wind data (Magnus Lindskog, HIRLAM personal communication WMO/WWRP Workshop on Use of NWP for Nowcasting, Boulder 2011 and HIRLAM Newsletter No. 58, November 2011).

In the USA, NCEP (National Center for Enviromental Prediction) has an operational RUC (Rapid Update Cycle) system with hourly data assimilation (Benjamin et al., 2004). As of September 2011, this was due to be replaced by the Rapid Refresh. The RR uses a version of the WRF model (currently v3.2+) and the Grid-point Statistical Interpolation (GSI) analysis largely developed at NCEP/EMC (Environmental Modelling Center, NOAA), using hourly cycling and a 13 km grid length. NCEP also run a 3 km model nested in the RR, but this has no separate data assimilation (Steve Weygandt et al., Earth System Research Lab, Boulder, personal communication WMO/WWRP Workshop on Use of NWP for Nowcasting, Boulder 2011, Stensrud et al., 2009; Smith et al., 2008; Weygandt et al., 2008).

In Japan, JMA (Japan Meteorological Agency) runs a Mesoscale Model (MSM) for Japan and its surrounding areas using a 5 km grid length and 4D-VAR with forecasts every 3 hours to 15 or 33 hours (Honda et al., 2005; Saito et al., 2006). This is a non-hydrostatic model (JMA-NHM). Development of NWP at a higher resolution (Local Forecast Model, LFM) is also in progress to help produce sophisticated disaster-prevention and aviation information services.

A trial operation of a 9-hour LFM forecast run on a 2 km grid length was performed in 2010 and 2011, and operational implementation is scheduled to start in 2012. LFM also uses JMA-NHM as a forecast model, and its initial condition is generated from a 3D-Var rapid update cycle. The cycle uses the MSM forecast as the first guess, and runs a JMA non-hydrostatic model-based variational data assimilation system (JNoVA) – a 3DVar (a degenerate version of JNoVA-4DVar) analysis and 1-hour JMA-NHM forecast in turn – over 3 hours using a 5 km grid length.

The Korean Meteorological Agency (KMA) is currently running the WRF 3D-Var (Barker et al., 2004, Xiao et al., 2008) at 10 km but is planning to use the 1.5 km, variable resolution Met Office Unified Model (UM) system with 3D-Var in the near future. In the past they have tested a 3.3 km version of WRF 4D-Var (Huang et al., 2009).

Over recent years, CAPS (Center for Analysis and Prediction of Storms, Oklahoma, USA) has been carrying out experimental, real time forecasting including the generation of 1 km grid length forecasts on a continental U.S. domain once a day, and the production of rapidly updated NWP-model-based nowcasts producing two hour, 1 km forecasts every 10 minutes (Xue et al., 2011; Kong et al., 2011; Clark et al., 2011; Brewster et al., 2010). These forecasts assimilate US operational WSR-88D radar data and/or high-resolution experimental X-band radar data, with and without assimilation cycles.

Comparison forecasts show systematically positive impacts of assimilating radar data on short-range precipitation forecasting, lasting up to 12 hours on average. To address forecast uncertainty and to provide probabilistic forecast information, storm-scale ensemble predictions have also been carried out and the products have been evaluated at an experimental forecasting facility. Extensive research has also been undertaken using ensemble-based data assimilation methods for initializing storm-scale NWP models, with very promising results.

CAPS has investigated a Mesoscale Convective System/vortex case study exploiting nested 400m /2 km grids and assimilating radar data at 5 min intervals using their Advanced Regional Prediction System (ARPS) 3DVAR+cloud analysis (Schenkman et al., 2011a) as well as EnKF (Snook et al., 2011). These show data impact on Collaborative Adaptive Sensing of the Atmosphere (CASA; Schenkman et al., 2011b) and probabilistic forecast skill with EnKF analyses (Snook et al., 2011; 2012).

Environment Canada has begun developing a convective-scale EnKF in order to examine the assimilation of radar data (e.g. over the Montreal region; Luc Fillion, personal communication WMO/WWRP Workshop on Use of NWP for Nowcasting, Boulder 2011). This is based on adaptation of the Global EnKF code available at Environment Canada (Houtekamer & Mitchell scheme) to a limited-area domain. The analysis step and the forecast model configuration (1 km horizontal grid length) are being validated.

5.4 Development of an NWP-based nowcasting system in the UK

5.4.1 Progress to date

The Met Office has run an operational, 4 km grid length NWP model for the UK (UK4) since December 2005. It has also run a 1.5 km UK configuration (UKV) routinely since summer 2010. Both models use three-hourly cycling 3D-Var and produce forecasts to 36 hours ahead with 70 levels. These are based on the Met Office's Unified Model (Davies et al., 2005) and variational data assimilation system (Lorenc et al., 2000; Rawlins et al., 2007), plus latent heat nudging (Macpherson et al., 1996; Jones & Macpherson, 1997; Dixon et al., 2009). They also include direct variational assimilation of analysed 3D cloud cover via associated relative humidity. The UKV has a 1.5 km grid length over the UK and a 4km stretched boundary nested in the 12 km NAE (North Atlantic and European) model. The UKV uses 3 km 3D-VAR over the whole domain. Collaborations with KMA and CAWCR (Centre for Australian Weather and Climate Research, Australian Government Bureau of Meteorology) are aiming to implement 1.5 km versions of the UM with 3D-Var or 4D-Var.

The Met Office's UK Post-Processing system (UKPP) incorporates a STEPS precipitation nowcast (Bowler et al., 2006). This combines a stochastic, radar-based extrapolation nowcast with UK4 or UKV precipitation forecasts. An 8 member ensemble and control member (unperturbed) nowcast to T+7 h are produced every 15 minutes. Recent Root Mean Squared Factor error statistics for STEPS control member advection nowcasts and UK4 and UKV forecasts of precipitation have shown that STEPS nowcasts are superior in the first 2.5 hours (see Figure 3).

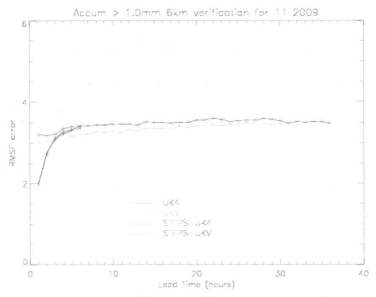

Fig. 3. Root Mean Squared Factor errors for November 2009 based on hourly accumulations greater than 1mm, smoothed to a scale of 6km, and measured using radar derived accumulations as the reference observation. The performance of 1.5 km (green) and 4 km (red) grid length, UK configurations of the Met Office's Unified Model are compared with control member STEPS nowcasts blending radar extrapolation with UM: 1.5 km (purple) and UM: 4km (blue) model forecasts. The performance of the STEPS nowcast blending extrapolation with the UM: 1.5 km forecast does not asymptote to that of UM: 1.5 km model because this was an experimental configuration run without prior calibration.

The implementation of an NWP-based nowcast system in the Met Office is focused on improving the prediction of convective storms for flood forecasting. The ultimate aim is to replace the existing extrapolation-based precipitation nowcasts and site specific forecasting techniques. Boundary conditions will be provided by the 6 hourly 1.5 km UKV system.

An hourly analysis and forecast system for southern England has been run experimentally for a limited number of cases of summer rain and convection, using conventional data and 3D-Var or 4D-Var, plus latent heat nudging of radar derived rain rates and humidity nudging based on analysed 3D cloud cover nudging (Macpherson et al., 1996; Jones & Macpherson, 1997; Dixon et al., 2009). The direct variational assimilation of cloud cover has not yet been tested in the hourly cycling system. This has used a fixed 1.5 km resolution configuration of the Unified Model and a 3 km resolution 4D-Var grid or 1.5 km and 3 km resolution 3D-Var grid.

The ultimate aim is to use 4D-VAR, if affordable and beneficial, in a real-time, routinely running NWP-based nowcast system. This will exploit high resolution (in time and space) Doppler radar measured radial winds and reflectivity or derived surface rain rates directly within the variational analysis scheme. Direct use in 4D-VAR should allow optimum extraction of information through interaction with other data sources, and the potential to modify the dynamical and physical forcing of precipitation and convective storms.

Research is also proceeding to investigate the background errors, balances and control variables required for use in convective scale data assimilation. The aim is to have a real-time system running continuously from summer 2012 for southern England.

5.4.2 A case study comparison of conventional and NWP-based nowcasts

Figure 4 compares T+1 hour, T+2 hour and T+3 hour STEPS control member nowcasts of surface precipitation rate, all valid at 2100 UTC on 3 June 2007 with a radar-based analysis of surface precipitation rate for the same time. At T+3 h, the STEPS nowcast is a combination of an extrapolation nowcast and UK4 forecast precipitation. The UK4 forecast tends to produce individual convective precipitation elements that are too large. It also fails to predict the full extent of the bands of convective precipitation to the east of the precipitation area lying through south-west England and west Wales. At T+2 hours, the STEPS scheme has re-produced the line of convection in the east but this is too narrow, possibly due to convergence in the diagnosed advection velocity field. By T+1 h, a reasonable nowcast has been produced.

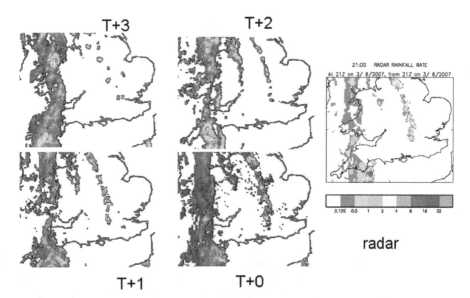

Fig. 4. STEPS T+0 h, T+1 h, T+2 h and T+3 h nowcasts of surface rain rate all valid at 2100 UTC on 3 June 2007. The key shown on the right-hand side represents precipitation rate in units of mm/h. Dry areas are shown in white. Note that dark blue areas in the STEPS nowcasts are not included in the colour key. These represent light drizzle.

Figure 5 shows the evolution of radar derived surface rain rate between 1200 UTC and 2100 UTC on 3 June 2007. It is apparent that a rain band in the west over Ireland at 1200 UTC reduces in intensity and moves only slightly eastward during the following 9 hours. However, bands to the east develop from about 1700 UTC onwards and intensify. The STEPS nowcast from 1800 UTC has not been able to reproduce the development of the eastern-most rain band seen in Figure 5. Nowcasts starting from later analysis times contain more precipitation but tend, incorrectly, to maintain the shape of individual features.

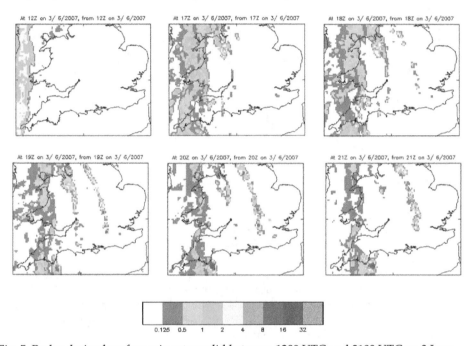

Fig. 5. Radar derived surface rain rates valid between 1200 UTC and 2100 UTC on 3 June 2007. The data for 1800 UTC, 1900 UTC, 2000 UTC and 2100 UTC were used to derive the T+3 h, T+2 h, T+1 h and T+0 h STEPS nowcasts shown in Fig. 4. The key shown below represents precipitation rate in units of mm/h. Note that dry areas are represented by the colour white.

Figure 6 compares T+1 hour, T+2 hour and T+3 hour 1.5 km NWP-based nowcasts of surface precipitation all valid at 2100 UTC on 3 June 2007 with radar derived precipitation rates for the same time. This model has used latent heat nudging of radar derived rain rates available every 15 minutes, and nudging of hourly humidity derived from 3-D cloud cover analyses in conjunction with hourly cycles of 4D-Var assimilation of conventional observations over 1 hour time windows. The NWP nowcasts improve at shorter lead times due to the benefit of data assimilation. In particular, they benefit from the latent heat nudging of surface precipitation rates derived from the sub-hourly radar data. In comparison with the STEPS nowcasts, the 1.5 km NWP nowcast has a better representation of the rain band in the east at both T+3 hours and T+2 hours. However, the representation of the rain in the south-west of England is inferior.

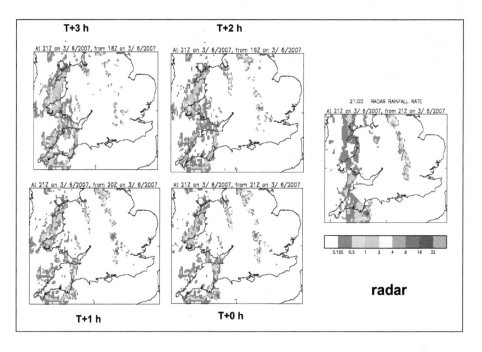

Fig. 6. A prototype Met Office NWP-based analysis (T+0 h) and T+1h, T+2h and T+3h nowcasts of surface rain rate generated using 4D-Var assimilation with latent heat nudging of the radar derived surface rain rates. All fields are valid at 2100 UTC on 3 June 2007. The area of coverage is the full domain of the prototype NWP-based nowcasting system. Note that dry areas are represented by the colour white.

Nonetheless, since this comparison is a first attempt without optimization of the data assimilation scheme and without the exploitation of more frequent conventional and Doppler radar measured radial wind observations, this is a very promising result. The forecast in the south-west can be improved by assimilation of 15 minute time frequency GPS water vapour data. At present, these are only available 90 minutes after data time so cannot be used in a nowcast system. Work is underway to make the UK GPS data available closer to data time.

Another potential source of water vapour information comes from radar refractivity by exploiting the interaction of the radar beam with ground clutter. Work is underway with Reading University to investigate the potential for obtaining this information from the UK weather radar network.

Direct assimilations of radar derived surface precipitation rates within 4D-Var is being investigated as well as direct or indirect assimilation of radar reflectivity, the latter through derived temperature and humidity increments from external 1D-Var assimilation of multiple beam elevations in vertical columns.

5.4.3 Use of Doppler radar derived winds

Potentially, weather radar provides a high resolution source of wind observations via the Doppler returns from hydro-meteors and insects. Currently, four weather radars in the south of England produce Doppler radial winds operationally every 5 minutes when there is precipitation (see Figure 7). The radars each perform scans at 5 elevations. The majority are at 1, 2, 4, 6 and 9 degrees, although one radar near London scans at 1, 2, 4, 5 and 5.5 degree elevations. Doppler winds are available to a range of about 100 km. This provides a small amount of dual or triple Doppler overlap in southern England as can be seen in Figure 7.

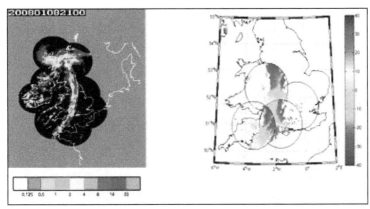

Fig. 7. A comparison of UK weather radar network coverage (left) and Doppler radar radial wind coverage (right) as of 8 January 2008. The key shown bottom-left represents precipitation rate in units of mm/hour. Note that dry areas in the left-hand graphic are represented by the colour black. The grey shading indicates areas without UK weather radar coverage.

Code has been developed to allow their processing, quality control, monitoring, super-obbing and data assimilation. Super-obbing is the process of combining observations that are of higher resolution than the forecast or analysis grid to reduce the data volume (Lorenc, 1981) and representativeness errors. Trials have been run to investigate the impact of Doppler radar radial winds on UK4 model forecasts using 3D-Var. The use of Doppler radial wind scans valid at analysis time was made operational in the UK configurations of the Unified Model in 2011. Three-hourly radial winds now replace hourly VAD winds from the same radars in the three-hourly 3D-Var cycles.

Much work has been done on specification of observation errors and investigating the impact of super-ob variances, errors derived from observation-background variances and errors derived from the Hollinsgworth and Lonnberg technique (Hollingsworth & Lonnberg, 1986).

The impact of Doppler radar radial wind data has been assessed over southern England using a prototype nowcasting system with a 1.5 km grid length model and hourly cycling 1.5 km 3D-VAR. For the initial tests, only the radar scans closest to the analysis hour were selected from each radar for assimilation. Initial subjective and objective verification looks promising. The location and coverage of precipitation is affected and improved in some situations. Figure 8 shows the increase in Fractional Skill Score (Roberts & Lean, 2008) of forecast hourly precipitation accumulations due to assimilation of radial winds from the four Doppler radars over southern England. These results are based on four case studies of about 10-19 cycles each, using hourly cycling 3D-Var and 11 hour forecasts. The results imply an hour's gain in skill in the earliest hours of the forecasts and a positive impact out to T+6 hours. The extent of the impact is limited by the small size of the domain and the spread of information from the boundary conditions into the domain.

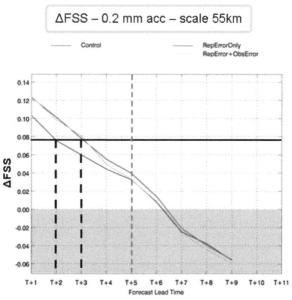

Fig. 8. ΔFSS for a 0.2 mm hourly precipitation accumulation threshold at a scale of 55km. Positive values of ΔFSS are indicative of forecast skill. The performance of the control forecast (blue) is compared with that of a forecast incorporating Doppler radial winds with a specified observation error derived from O-B statistics and referred to as representativeness error (red), and a similar forecast including Doppler radial winds with the representativeness error plus the super-observation standard deviation as the observation error (green).

Work continues on the specification of observation error and to test the impact of hourly and higher time frequency data in the 4D-Var prototype nowcasting system. The impact and areal influence of observations in a NWP analysis depends on the background error correlation and covariances (i.e. the short range forecast error) at the analysis time, in addition to the observation error itself. The background errors can have a significant impact on forecast quality and the benefit afforded by the observations. Thus, work is underway to define improved errors for the 1.5 km grid length forecasts, both in terms of correlations between variables, length scales and error variances. These need to extract longer time and synoptic scale information as well as information at shorter time and spatial scales from radar data with high spatial and time resolutions. This is very challenging work.

Work with Reading University has been undertaken to look at the potential for use of winds derived from insect returns in fine weather (Rennie et al., 2010). This will continue in collaboration with CAWCR in Australia. Radar returns only give radial winds (i.e. in the direction of the radar beam) rather than 3-D wind components, so the additional information in areas of overlapping radars (dual-Doppler) may increase the impact of wind retrievals in those locations.

5.4.4 Conclusions and further work

1.5 km grid length NWP in the Met Office is showing promise in the very short range prediction of convection over the UK. Previous sections have highlighted the potential benefits of using radar derived precipitation rates through latent heat nudging on top of 4D-Var and of using Doppler radar derived radial winds in 3D-Var.

4D-Var has the potential to exploit higher time frequency observations and to extract more information from them than 3D-Var. Therefore, research is continuing on the use of high time frequency Doppler radial winds, direct application of radar derived surface precipitation rate, and direct and indirect use of multi-elevation volume scan reflectivity in 4D-Var. Although latent heat nudging is still showing benefit in forecasts, it cannot correctly represent resolved convection where latent heat release occurs in different locations to surface precipitation, so it is hoped to obtain benefits from direct 4D-Var or indirect 1D-Var assimilation of the reflectivity data.

Unfortunately, 4D-Var is computationally expensive on the super-computer currently available to the Met Office. Therefore, research and development is being undertaken with both 3D-Var and 4D-Var systems. With a super-computer upgrade due in 2012, the aim is to start running a prototype real-time NWP-based nowcast system in 2012, hopefully with 4D-Var if the upgrade provides sufficient computer resources.

Due to the tight time constraints imposed by operational schedules, it may be necessary to move away from use of a time window centred on the analysis time to one finishing at the analysis time. High quality data sources such as GPS, which provide information on low level humidity, are currently only available 90 minutes after data time, although less accurate but more timely data may become available. There are many sources of information on different variables (e.g. GPS, radar refractivity, satellite imagery and surface observations for low level humidity). The usefulness of the different data sources will be investigated to provide an optimum system. The initial experiments reported here were undertaken nested within the UK 4 km NWP forecast system. Now, the nowcasting system is being tested embedded in the UK 1.5 km NWP forecast system.

The skill of the convective scale nowcasts is very dependent on the accuracy of the synoptic forcing conditions both within the nowcast domain itself and the boundary conditions. Both the UK models and the embedded nowcast system use the same model and essentially the same data assimilation system. Errors in convective initiation can come from errors in the synoptic flow either as a result of lack of observations to correct model errors, or incorrect or sub-optimal use of observations. Finding the best way to extract synoptic scale and convective scale information from observations in both the nowcast system itself and in the forcing at the boundaries will be key to improvements in the skill of the nowcast.

Data sources such as GPS can be problematical because they are vertically integrated measurements depending on the accuracy of the specification of the forecast background errors, and interaction with other data sources to allocate changes to humidity in the vertical can have dramatic impacts on forecast precipitation. The use of high vertical and horizontal spatial and temporal resolution Doppler radar winds and reflectivity or rain rate data, and improvements in specification of forecast background errors, can lead to changes and improvements in the impact of different data sources and the accuracy of the precipitation in the early hours of the forecast.

NWP systems can suffer from imbalances in the initial conditions leading to spin-up or spin-down of precipitation in the initial stages of the forecast. Work to improve this will help to improve skill in the early hours of the nowcasts. Improvements in the skill of the forecast model itself in terms of precipitation biases are likely to help both the forecast and the ability to assimilate observations. We tend to use radar derived rain rates to verify the NWP forecasts, for assimilate into the models and to improve the formulation of the model. However, the radar data can have quality issues, for example relating to attenuation, and improvements in quality control and data processing are needed to ensure that the radar data are of high quality.

The entire UK network of weather radars will gradually be updated to produce Doppler radial winds and also dual-polarization data and radar refractivity measurements. The use of radar data in NWP high resolution variational data assimilation has the potential to improve on current extrapolation-based nowcasts. To achieve this we need high quality radar data, fast processing (techniques and computer power), careful specification of observation and forecast background error covariances and correlations through the scientific design of the data assimilation system, and a good representation of the dynamical and microphysical processes in the NWP forecast model.

In future it is hoped to exploit ensemble techniques in both the data assimilation and production of forecasts. If there is sufficient computer power available for hourly NWP forecasts to 12 hours, this will provide the potential for 6 hours of 1 hourly lagged ensemble forecasts and a measure of the predictability of the nowcasts.

6. Application of radar-based precipitation nowcasts to hydrological forecasting and warning

6.1 Overview

Documented uses of radar data in hydro-meteorology are many and varied. They include numerous studies of the space-time structure of radar inferred precipitation fields (e.g. Harris et al., 2001), the compilation of precipitation climatologies (Panziera et al., 2011), the estimation of Probable Maximum Precipitation (Cluckie, Pessoa & Yu, 1991; Collier & Hardaker, 1996), reservoir design and safety (Cluckie & Pessoa, 1988), design storm modelling (Seed, 2003), urban drainage and waste water management (Cluckie & Tyson, 1989; Schellart et al., 2009), river flow management (Lewin, 1986) and hydroelectric power generation (Baker, 1986).

In addition to the above, operational radar-based precipitation nowcasts can be of great value in fluvial (river) flood prediction because they extend the lead time of flood warnings by reducing reliance on crude assumptions regarding future precipitation. For pluvial (surface water) flood forecasting, predictions of future precipitation are essential because the time between the precipitation reaching the ground and any consequent flooding is very short (Golding, 2009). In this section we review some of the key developments in the use of radar for fluvial (river) flood prediction and warning.

6.2 Hydrological requirements for precipitation observations

It was the prospect of accurate, contiguous observations of precipitation over large areas that first stimulated hydrologists to explore the use of radar data for the prediction of run-off and river flow. Early assessments of the value of radar data were mixed (Anderl et al., 1976; Barge et al., 1979). This is not surprising given the reliance of these early experiments on deterministic precipitation estimates of variable accuracy, and the many factors known to impact on hydrological forecast performance.

Hydrological requirements for precipitation observations and forecasts are a function of catchment size, morphology and land use, and the hydrological model used (Hudlow et al., 1981). Many operational, hydrological forecasting models are lumped conceptual models in which the catchment response is modelled as a whole and the precipitation input is an areal average estimate. A number of authors have emphasized that the benefits of radar derived, spatially contiguous precipitation estimates can only be fully realized if used as input to distributed, conceptual or physically-based hydrological models (e.g. Moore, 1987).

6.3 Impact of the spatial and temporal distribution of precipitation

Wilson et al. (1979) found that a failure to properly represent the spatial distribution of rainfall, due to reliance on point observations from rain gauges, could produce significant errors in the total volume, peak and time to peak of an estimated hydrograph, even when the rainfall depth and its temporal evolution were accurately recorded at rain gauge sites. Errors were largest in cases of localized convective storms. Bedient and Springer (1979) demonstrated that the peak flows in a catchment could be enhanced when the precipitation moved in the direction of the stream.

More recently, Bell and Moore (2000b) explored the sensitivity of lumped and distributed catchment rainfall–run-off models to time series of rainfall observations from radar and rain gauge, gridded to a range of spatial resolutions. For a small rural catchment, they confirmed the sensitivity of distributed model run-off to the spatial variability of rainfall. A comparison of the performances of lumped and distributed models showed similar levels of predictive skill in stratiform rain, but superior distributed model predictions during convective rainfall events.

Ball (1994) examined the impact of the temporal evolution of the precipitation pattern on the time of concentration and peak discharge in a catchment. The time of concentration was shown to be sensitive to the temporal evolution of excess rainfall over the catchment, where as catchment peak discharge was not. Thus, timing errors in predicted flows can result if the time interval between precipitation observations is too long. Collier (1996) suggests that a radar scan cycle of no more than 5 minutes is required to capture the time evolution of most convective precipitation fields.

6.4 Relationship between radar data resolution, catchment characteristics and hydrological model performance

Various studies have examined the impact of the spatial and temporal resolution of remotely sensed precipitation observations on hydrological model performance (Krajewski et al., 1991; Pessoa et al., 1993; Obled et al., 1994; Ogden & Julien, 1994; Ball, 1994; Faurès et al,, 1995; Shah et al., 1996; Winchell et al., 1998; Bell & Moore, 2000b; Carpenter et al., 2001).

Ogden and Julian (1994) explored the relationship between catchment size, and the correlation length and horizontal resolution of the radar derived precipitation fields input to a two-dimensional, physically-based hydrological model. They defined two, dimensionless length parameters and considered their impacts on the accuracy of predicted run-off. *Storm smearing* describes a reduction in the horizontal gradient of precipitation rate as the horizontal resolution of the radar data approaches its correlation length. *Watershed smearing* occurs when the horizontal resolution of the radar data approaches the characteristic length scale of the catchment (square root of the catchment area). Watershed smearing was shown to be the main source of error in predicting river flow over small catchments. Berenguer et al. (2005) point out that the sensitivity of hydrological models to biases in mean areal rainfall are due to the fact that river catchments act as integrators of the precipitation falling on them.

Bell and Moore (2000b) emphasized the need to calibrate hydrological models with rainfall data for a given resolution. They found that the most skilful distributed rainfall-run-off model predictions were made with lower resolution rainfall data. This finding was interpreted as evidence for the need to improve distributed hydrological model formulation.

6.5 Impact of radar intensity resolution

The impact of the intensity resolution in radar data on hydrological forecast errors was investigated by Cluckie, Tilford & Shepherd (1991). They demonstrated that 8 intensity levels were adequate for the majority of rural and urban catchments in the majority of UK precipitation events. This is because the bulk of the relevant information content is concentrated at the low frequency end of the power spectrum. Nonetheless, in convective precipitation events, a reduction in intensity resolution may have an effect similar to that of the storm smearing described by Ogden and Julian (1994).

6.6 Benefits of precipitation nowcasts to hydrological forecasting

In the absence of precipitation forecasts, the lead time of flood warnings is limited by the catchment response time, a quantity dependent on catchment size, morphology and land use. Skilful precipitation forecasts offer the prospect of some forewarning of flash floods in small, fast responding catchments, and of extending the lead time of flood warnings in other catchments (Roberts et al., 2009).

Although numerous authors have evaluated the impact of QPE algorithms on the utility of radar for hydrological forecasting there have been relatively few investigations of the benefits of precipitation nowcasts in this area. Cluckie and Owens (1987) compared the performance of stream flow forecasts made using a linear transfer function model and radar extrapolation nowcasts from FRONTIERS (Browning, 1979) against similar flow predictions made using average past rainfall and an assumption of no more rain. In most cases, they found that nowcast driven hydrological forecasts outperformed the alternatives, although on one occasion they showed the former to be poor.

Several decades later, a similar, case study orientated evalution of the utility of Nimrod (Golding, 1998) extrapolation nowcasts for rainfall-run-off modelling in Scotland (Werner and Cranston, 2009) drew similar conclusions: although errors in nowcast driven predictions of river flows could be substantial, they were smaller than those of flow forecasts made assuming zero future rainfall.

Mecklenburg et al. (2001) found that COTREC-based radar extrapolation nowcasts (Lagrangian persistence) produced superior hydrological forecasts to Eulerian persistence using a lumped conceptual model. In a similar vein, Berenguer et al. (2005) compared hydrological forecasts made with the S-PROG model (Seed, 2003) with those produced using a simpler, extrapolation-based precipitation nowcast in a Mediterranean environment. S-PROG utilizes a scale decomposition framework and associated hierarchy of auto-regressive models to smooth the advected precipitation field at a rate that is consistent with its loss of predictive skill on a hierarchy of scales. This approach is intended to minimize the root mean squared forecast error. Berenguer et al. (2005) concluded that radar-based precipitation nowcasts in general could extend the lead time of useful hydrological forecasts from 10 minutes to over an hour in a fast response responding Mediterranean catchment. However, the results obtained with S-PROG were not significantly better than those obtained with a simpler Lagrangian persistence technique.

Since one of the key benefits of radar is its ability to provide contiguous, instantaneous observations of precipitation over a wide area, other studies have focused their efforts on demonstrating the benefits of precipitation nowcasts when input to distributed hydrological models. In these models, the run-off response can vary within a catchment according to the temporal and spatial variability of the rainfall, surface properties and antecedent wetness (Ivanov et al., 2004; Vivoni et al., 2005). Amongst other things, this capability allows time series of run-off to be generated at ungauged sites (Moore et al., 2007).

Sharif et al. (2006) explored the potential of the National Center for Atmospheric Research's Auto-Nowcaster to improve the lead time and accuracy of hydrological forecasts made with a physically-based distributed parameter model. Rain gauge and radar observation driven simulations were used as a baseline. Results confirmed that the use of precipitation nowcasts could significantly improve flood warning in urban catchments, even in the case of short-lived events in small catchments. Similar conclusions were drawn by Vivoni et al. (2006) in relation a set of small, mixed land-use catchments in Oklahoma, in this case using NEXRAD-based extrapolation nowcasts and a distributed hydrological model.

6.7 Treatment of nowcast errors in hydrological forecasts

Vivoni et al. (2007) explored the impact of errors in deterministic precipitation nowcasts on errors in flood forecasts using a distributed hydrological model and a range of catchment sizes. Their investigations showed that increases in nowcast error with lead time produced larger errors in the resulting hydrological forecasts. They demonstrated that the effects of nowcast errors could be simultaneously enhanced or dampened in different locations depending on forecast lead time and precipitation characteristics. Differences in error propagation between sub-catchments were effectively averaged out over larger catchment areas.

Despite continuing incremental advances in radar technology and performance during the 1970s and 1980s, a number of authors recognized that radar derived estimates of surface precipitation rate and accumulation would remain subject to hydrologically significant errors, particularly in hilly and mountainous areas. Collier and Knowles (1986) concluded that the full benefits of radar to operational hydrology would only be realized when ways could be found of accounting for these errors.

This thinking coincided with a growing awareness of the need to develop operational systems integrating meteorological and hydrological forecast models (Georgakakos and Kavvas, 1987). Early examples of such systems are the Integrated Flood Observing and Warning System (IFLOWS) implemented in the USA (Barrett and Monro, 1981), and the Regional Communication Scheme in north-west England, integrating operational weather radar data with hydrological forecasting and warning under the North-West radar project (Noonan, 1987).

More recently, a number of multi-national initiatives including HEPEX (e.g. Buizza, 2008; Pappenberger et al., 2008), MAP-D-Phase (Zappa et al., 2008; Bogner & Calas, 2008) and COST-731 (Rossa et al., 2011) have supported work to implement integrated flood forecasting systems exploiting weather radar. A number of UK-based research programmes, are relevant in this context, including HYREX (e.g. Mellor et al., 2000a,b; Bell & Moore, 2000a), the Natural Environment Research Council's Flood Risk from Extreme Events (FREE), the Engineering and Physical Sciences Research Council's Flood Risk Management Research Consortium (FRMRC) and FLOODsite.

Until Krzysztofowicz's pioneering work (Krzysztofowicz, 1983, 1993, 1998, 1999, 2001) to develop and implement an integrated hydro-meteorological systems framework, incorporating a Bayesian treatment of uncertainties in data inputs and deterministic model forecasts, errors in the precipitation inputs to operational hydrological models tended to be handled through the use of what-if scenarios (Haggett, 1986; Werner et al., 2009). The derived distribution approach (Seo et al., 2000) developed by Krzysztofowicz exploits the total probability law to derive by quasi-analytic means the conditional probability distribution of river stage given the initial and boundary conditions, including future precipitation parameterized in the form of a probabilistic QPF.

An alternative method for accounting for data input uncertainty entails the ingestion of ensemble precipitation forecasts into a hydrological model, whilst taking separate account of other sources of uncertainty contributing to the total uncertainty in the hydrological forecast variable of interest (Krzysztofowicz, 2001). Probabilities of exceeding specific river stage thresholds can then be estimated from the resulting ensemble of hydrographs (Schaake & Larson, 1998; Pierce et al., 2004). During the past decade, this ensemble approach has gained credence with the widespread implementation of operational ensemble NWP models and the development of stochastic nowcasting and post-processing techniques for the production of ensembles of high resolution precipitation nowcasts (e.g. Bowler et al., 2006).

Carpenter and Georgakakos (2006) investigated the combined effects of radar rainfall errors and catchment size on the uncertainty in predicted river flow with the aid of a distributed hydrological model. Using a parsimonious model to represent the spatial structure and variance of radar errors, they demonstrated that ensemble spread in predicted flow was log-linearly related to catchment scale.

A handful of ensemble-based probabilistic QPN schemes have been developed during the past decade or so. These were described earlier in this Chapter and include the String of Beads Model (Pegram & Clothier, 2001; Berenguer et al., 2011), the Short-Term Ensemble Prediction System (Bowler et al., 2006) and a method recently described by Kober et al. (2011). Here we draw a distinction between ensemble QPN schemes and others such as those described by Andersson and Iverson (1991) and Germann and Zawadzki (2004) which produce forecasts of the probability distribution of precipitation at a point. The latter cannot be used for ensemble-based probabilistic hydrological forecasting because they do not provide a complete description of the joint probability distribution of precipitation, which plays a key role in the hydrological response of a catchment.

In the UK, the Department of the Environment, Fisheries and Rural Affairs and the Environment Agency recently funded an R&D project to explore the benefits of high resolution precipitation forecasts to fluvial flood prediction and warning (Schellekens et al., 2010). The potential for operational use of ensemble rainfall nowcasts from STEPS (Bowler et al., 2006) was investigated in conjunction with lumped and distributed rainfall–run-off models. The evaluation included hydrological configuration issues, data volumes, run times and options for displaying probabilistic forecasts. No quantitative verification of the precipitation ensemble-driven hydrological forecasts was undertaken.

Recently, the Environment Agency has implemented a nationally configured, distributed hydrological model, known as Grid-to-Grid (Bell et al., 2007). In the near future (2012), this model will be driven by ensemble precipitation forecasts integrating STEPS ensemble nowcasts.

7. The future of nowcasting

One of the major changes in the past decade has been the increase in the ability of the general population in developed countries to receive real-time information over a range of mobile platforms. This makes it possible to deliver location specific nowcasts to millions of users. They use this information to make routine decisions regarding leisure and other outdoor activities and very occasionally decisions relating to severe weather events. Mitigating damage due to severe weather has been the motivation for developing nowcasting systems in the past, but the focus is likely to change to providing routine nowcasting services to the general public. The current generation of nowcasting systems are already capable of delivering products that are useful in this context and the focus in the short term should be on developing the ability to customize and disseminate these to a very large number of users.

Improved communications and computer capacity have also made it possible to routinely combine data from a network of weather radars into a single large domain, and to improve the algorithms that are used to provide quantitative radar rainfall estimates. Improvements in the QPE will continue as the radar hardware improves and the density of the radar networks increases. These improvements will allow for improvements in the quality of the nowcasts in the first hour.

There are still incremental gains to be made by improving the accuracy of the tracking algorithms and by combining the cell tracking and field advection paradigms into a single advection scheme. More generally, there is evidence that the cell tracking and field tracking systems are complementary, so rather than viewing them as competitors there should be value in developing a way of optimally combining the forecasts from several nowcasting systems based on an analysis of which system is likely to be providing better nowcasts in any given situation.

Predicting the initiation and decay of convective storms will continue to be a major focus for research because gains in this area will lead to significant improvements in the accuracy of nowcasts beyond 30 minutes. The problem with heuristic and analogue techniques is that they require large data sets for calibration, and the associated conceptual models that are developed tend to be location specific. This can be overcome if a way can be found to allow the algorithms to learn as they go, based on the results of routine real-time verification.

Possibly the major use of radar data in the future will be for assimilation into NWP models of the national weather services that run radar networks. Empirical advection nowcasting will continue to provide nowcasts, but for more limited lead times as the NWP models gain accuracy at shorter lead times. Not all users will be able to afford the costs of a full NWP system that is able to assimilate radar data and there will continue to be a demand for fast and cheap rainfall nowcasts for specific purposes.

Forecast errors, rather like death and taxes, will always be with us and the future lies in using ensembles or other techniques to convey the uncertainty in the current forecast to the users. Further research on quantifying forecast errors and understanding how they depend on location and meteorological situation is required before we are able to demonstrate that the spread in a nowcast ensemble fully represents the uncertainty. There is also a need to develop probabilistic nowcasting systems that do not only forecast rainfall, but are used to forecast end-user impacts, for example the traffic capacity of an air-corridor, or the water level in a river.

8. References

Anagnostou, E. N., Krajewski, W. F. & Smith, J. A. (1999). Uncertainty quantification of mean-areal radar-rainfall estimates, *Journal of Atmopheric and Oceanic Technology*, Vol. 16, No. 2, (February 1999), pp. 206–215, Available from http://dx.doi.org/10.1175/1520-0426(1999)016<0206:UQOMAR>2.0.CO;2

Anderl, B., Attmannspacher, W. & Schultz, G. A. (1976). Accuracy of reservoir inflow forecasts based on radar rainfall measurements, *Water Resources Research*, Vol. 12, No. 2, pp. 217-223, doi:10.1029/WR012i002p00217

Andersson, T. & Ivarsson, K.-I. (1991). A model for probability nowcasts of accumulated precipitation using radar. *Journal of Applied Meteorology*, Vol. 30, (January 1991), pp. 135–141

Austin, P. M. (1987). Relation between measured radar reflectivity and surface rainfall. *Monthly Weather Review*, Vol. 115, No. 5, pp. 1053–1070, ISSN 00270644

Austin, G. L. & Bellon, A. (1974). The use of digital weather records for short-term precipitation forecasting, *Quarterly Journal of the Royal Meteorological Society*, Vol. 100, No. 426, (), pp. 658-664

Austin, G. L. & Bellon, A. (1982). Very short-range forecasting of precipitation by the objective extrapolation of radar and satellite data, In: *Nowcasting*, K. A. Browning (Ed.), 177-190, Academic Press, London, UK

Baker, S. E. (1986). The relationship of QPF to the management of hydroelectic power on the Santee River Basin in North and South Carolina, *Proceedings of the Conference on Climate and Water Management: A Critical Era*, pp. 77-82, Asheville, North Carolina, American Meteorological Society, Boston, Massachusetts, 4-7 August 1986

Ball, J. E. (1994). The influence of storm temporal patterns on catchment response, *Journal of Hydrology*, Vol. 158, NO. 3-4, pp. 285-303, ISSN 0022-1694

Ballard, S., Zhihong, L., Simonin, D., Buttery, H., Charlton-Perez, C., Gaussiat, N. & Hawkness-Smith, L. (2011). Use of radar data in NWP-based nowcasting in the Met Office, *Proceedings of the eighth International Symposium on Weather Radar and Hydrology*, Exeter, UK, April 2011, to appear in IAHS red book

Bally, J. (2004). The Thunderstorm Interactive Forecast System: Turning automated thunderstorm tracks into severe weather warnings, *Weather Forecasting*, Vol. 19, No. 1, (February 2004), pp. 64–72

Barge, B. L., Humphries, R. G., Mah, S. J. & Kuhnke, W. K. (1979). Rainfall measurements by weather radar: applications to hydrology, *Water Resources Re*search, Vol. 15, No. 6, pp. 1380-1386, doi:10.1029/WR015i006p01380

Barker, D.M., Huang, W., Guo, Y.-R. & Xiao, Q. (2004). A three-dimensional variational (3DVAR) data assimilation system for use with MM5: Implementation and initial results, *Monthly Weather Rev*iew, Vol. 132, pp. 897-914

Barrett, C. B. & Monro, J. C. (1981). National prototype flash flood warning system. Preprint Volume, *4th Conference on Hydrometeorology*, pp. 234-239, Reno, Nevada, 7-9 October 1981

Battan, L. J. (1953). Observations of the formation and spread of precipitation in cumulus clouds, *Journal of Meteorol*ogy, Vol. 10, pp. 311-324

Bedient, P. B. & Springer, N. K. (1979). Effect of rainfall timing on design floods, *Journal of Civil Engineering Desi*gn, Vol. 1, No. 4, pp. 311-323

Bell, V. A., Kay, A. L., Jones, R. G. & Moore, R. J. (2007). Development of a high resolution grid-based river flow model for use with regional climate model output, *Hydrology and Earth System Sciences*, Vol. 11, No. 1, pp. 532-549, doi:10.5194

Bell, V.A. & Moore, R.J. (2000a). Short period forecasting of catchment-scale precipitation. Part II: a water-balance storm model for short-term rainfall and flood forecasting, *Hydrology and Earth System Sciences*, Vol. 4, No. 4, pp. 635–651

Bell, V.A. & Moore, R.J. (2000b). The sensitivity of catchment runoff models to rainfall data at different spatial scales, *Hydrology and Earth System Sciences*. Vol. 4, No. 4, pp. 653-667

Bellon, A. & Austin, G. L. (1978). The evaluation of two years of real time operation of a short-term precipitation forecasting procedure (SHARP), *Journal of Applied Meteorol*ogy, Vol. 17, No. 12, pp. 1778–1787, DOI: 10.1175/1520-0450

Bellon, A. & Austin, G. L. (1984). The accuracy of short-term radar rainfall forecasts, *Journal of Hydrology*, Vol. 70, Nos. 1-4, (February 1984), pp. 35-49

Benjamin, S. G., Schwartz, B. E., Szoke, E. J., & Koch, S. E. (2004). The value of wind profiler data in U.S. weather forecasting, *Bulletin of the American Meteorological Society*, Vol. 85, pp. 1871-1886

Berenguer, M., Corral, C., Sanchez-Diezma, R. & Sempere-Torres, D. (2005). Hydrological validation of a radar-based nowcasting technique, *Journal of Hydrometeorology*, Vol. 6, No. 4, pp. 532-549

Berenguer, M., Sempere-Torres, D. & Pegram, G. G. S. (2011). SBMcast – An ensemble nowcasting technique to assess the uncertainty in rainfall forecasts by Lagrangian extrapolation, *Journal of Hydrology*, Vol. 404, pp. 226-240

Blackmer, R. H., Duda, R. O. & Reboh, R. (1973). *Application of pattern recognition to digitized Weather Radar Data*. Final Rep., Contract 1-36072, SRI Project 1287, Stanford Research Institute, Menlo Park, CA, 89 pp., Available from National Information Service, Operations Division, Springfield, VA 22161.

Bogner, K. & Kalas, M. (2008). Error-correction methods and evaluation of an ensemble based hydrological forecasting system for the Upper Danube catchment. *Atmospheric Science Letters*, Vol. 9, No. 2, (April 2008), pp. 95–102, DOI: 10.1002

Bowen, E. G. (1951). Radar observations of rain and their relation to mechanisms of rain formation. *Journal of Atmospheric and Terrestrial Physics*, Vol. 1, pp. 125-140

Bowler, N. E., Pierce, C. E. & Seed, A. W. (2004). Development of a precipitation nowcasting algorithm based upon optical flow techniques, *Journal of Hydrology*, Vol. 288, pp. 74–91

Bowler, N. E., Pierce, C. E. & Seed, A. W. (2006). STEPS: A probabilistic precipitation forecasting scheme which merges an extrapolation nowcast with downscaled NWP, *Quarterly Journal of the Royal Meteorological Society*, Vol. 132, pp. 2127-2155

Brewster, K., Thomas, K. W., Gao, J., Brotzgel, J., Xue, M. & Wang, Y. (2010). A nowcasting system using full physics numerical weather prediction initialized with CASA and Nexrad radar data, *Proceedings of the 25th Conference on Severe Local Storms*, pp. 11, American Meteorological Society, Denver, Colorado, October 11-14

Brousseau, P., Berre, L., Bouttier, F. & Desroziers, G. (2011). Background-error covariances for a convective-scale data-assimilation system : AROME-France 3D-Var, *Quarterly Journal of the Royal Meteorological Society*, Vol. 137, No. 655, pp. 409-422

Browne, I. C. & Robinson, N. P. (1952). Cross-polarization of the radar melting band, *Nature*, Vol. 170, (December 1952), pp. 1078-1079

Browning, K. A. (1979). The FRONTIERS plan: a strategy for using radar and satellite imagery for very short-range precipitation forecasting, *Meteorological Magazine*, Vol. 108, pp. 161-184

Browning, K. A. (1980). Local weather forecasting, *Proceedings of the Royal Society of London*, Series A, Vol. 371, No. 1745, pp. 179–211

Buizza, R. (2008). The value of probabilistic prediction, *Atmospheric Science Letters*, Vol. 9, No. 2, pp. 36–42

Byers, H. R. (1948). The use of radar in determining the amount of rain falling over a small area, *Eos Transactions of the American Geophysical Union*, Vol. 29, pp. 187-196

Byers, H. R. & Braham Jr., R. R. (1949). *The Thunderstorm*. U.S. Govt. Printing Office, 187 pp

Carpenter, T. M. & Georgakakos, K. P. (2006). Discretization scale dependencies of the ensemble flow range versus catchment area relationship in distributed hydrologic modelling, *Journal of Hydrology*, Vol. 328, pp. 242-257

Carpenter, T. M., Georgakakos, K. P. & Sperfslage, J. A. (2001). On the parametric and NEXRAD-radar sensitivities of a distributed hydrologic model suitable for operational use, *Journal of Hydrology*, Vol. 253, pp. 169-193

Caumont, O., Ducrocq, V., Wattrelot, E., Jaubert, G. & Pradie-vabre, S. (2010). 1D+3DVar assimilation of radar reflectivity data: a proof of concept, *Tellus A*, Vol. 62, pp. 173-187

Caya, A., Sun, J. & Snyder, C. (2005). A comparison between the 4D-Var and the ensemble Kalman filter techniques for radar data assimilation, *Monthly Weather Review*, Vol. 133, No. 11, pp. 3081-3094, ISSN 0027-0644

Ciach, G. J., Krajewski, W. F. & Villarini, G. (2007). Product-error-driven uncertainty model of probabilistic quantitative precipitation estimation with NEXRAD data, *Journal of Hydrometeorology*, Vol. 8, pp. 1325-1347

Clark, A. J., Weiss, S. J., Kain, J. S., Jirak, I. L., Coniglio, M., Melick, C. J., Siewert, C. , Sobash, R. A., Marsh, P. T., Dean, A. R., Xue, M., Kong, F., Thomas, K. W., Wang, Y., Brewster, K., Gao, J., Wang, X., Du, J., Novak, D. R., Barthold, F. E., Bodner, M. J., Levit, J. J., Entwistle, C. B., Jensen, T. L. & Correia, J. Jn. (2011). An Overview of the 2010 hazardous weather testbed experimental forecast program spring experiment, *Bulletin of the American Meteorological Society*, doi: 10.1175/BAMS-D-11-00040.1

Cluckie, I.D. & Owens, M. D. (1987). Real-time rainfall-runoff models and use of weather radar information, In: *Weather Radar and Flood Forecasting*, V. Collinge & C. Kirby (Eds.), 171-190, John Wiley & Sons, Chichester

Cluckie, I. D. & Pessoa, M. A. (1988). Weather radar and dam safety: An evaluation, *Proceedings of the 2nd Anglo/Polish Hydrology Colloquium*, held at the University of Birmingham under the auspices of the Polish Academy of Sciences and the Royal Society of London, August 1988

Cluckie, I. D., Pessoa, M. L. & Yu, P. S. (1991). Probable maximum flood modelling utilising transposed, maximised radar-derived precipitation data, In: *Hydrological Applications of Weather Radar*, I. D. Cluckie & C. G. Collier (Eds.), 181-191, Ellis Horwood, London

Cluckie, I. D., Tilford, K. A. & Shepherd, G. W. (1991). Radar signal quantization and its influence on rainfall-runoff models, In: *Hydrological Applications of Weather Radar* , I. D. Cluckie & C. G. Collier (Eds.), 440-451, Ellis Horwood, London

Cluckie, I. D. & Tyson, J. M. (1989). Weather radar and urban drainage systems. In: *Weather Radar and the Water Industry: Opportunities for the 1990s*, 65-73, BHS Occasional Paper No. 2, British Hydrological Society

Collier, C. G. (1996). *Applications of Weather Radar Systems: a guide to uses of radar data in meteorology and hydrology*, John Wiley and Sons Ltd

Collier, C. G. & Hardaker, P. J. (1996). Estimating Probable Maximum Precipitation using a storm model approach, *Journal of Hydrology*, Vol. 183, pp. 277-306

Collier, C. G. & Knowles, J. M. (1986). Accuracy of rainfall estimates by radar, Part III: Application for short-term flood forecasting, *Journal of Hydrology*, Vol. 83, pp. 237-249

Cox, D. R. & Isham, V. (1988). A simple spatio-temporal model of rainfall, *Proceedings of the Royal Society of London*, Series A, Vol. 415, No. 1849, pp. 317-328

Crane, R. K. (1979). Automatic cell detection and tracking, *Institute of Electrical and Electronics Engineers (IEEE) Transactions on Geosciences and Electronics (GE)*, Vol. 17, No. 4, pp. 250-262

Dance, S., Ebert, E. E. & Scurrah, D. (2010). Thunderstorm strike probability nowcasting, *Journal of Atmospheric and Oceanic Technology*, Vol. 27, pp. 79–93

Davies, T., Cullen, M. J. P., Malcolm, A. J., Mawson, M. H., Staniforth, A., White, A. A. & Wood, N. (2005). A new dynamical core for the Met Office's global and regional modelling of the atmosphere, *Quarterly Journal of the Royal Meteorological Society*, Vol. 131, pp. 1759-1782

Dixon, M. & Wiener, G. (1993). TITAN: Thunderstorm Identification, Tracking, Analysis and Nowcasting — A radar-based methodology, *Journal of Atmospheric and Oceanic Technology*, Vol. 10, pp. 785–797

Dixon, M., Li, Z., Lean, H., Roberts, N. & Ballard, S. P. (2009). Impact of data assimilation on forecasting convection over the United Kingdom using a high-resolution version of the Met Office Unified Model, *Monthly Weather Review*, Vol. 137, pp. 1562–1584

Duda, R. O. & Blackmer, R. H. (1972). *Applications of pattern recognition techniques to digitized weather radar data*, Final Rep. Contract I-36092, SRI Project 1287, Stanford Research Institute, Menlo Park, CA, 135 pp., Available from National Information Service, Operations Division, Springfield, VA 22161

Ebert, E. E., Wilson, L. J., Brown, B. G., Nurmi, P., Brooks, H. E., Bally, J. & Jaeneke, M. (2004). Verification of nowcasts from WWRP Sydney 2000 Forecast Demonstration Project, *Weather and Forecasting*, Vol. 19, pp. 73 – 96

Fabry, F. & Seed, A. W. (2009). Quantifying and predicting the accuracy of radar-based quantitative precipitation forecasts, *Advances in Water Resources*, Vol. 32, No. 7, pp. 1043-1049

Faurès, J. M., Goodrich, D. C., Woodlhiser, D. A. & Sorooshian, S. (1995). Impact of small-scale spatial rainfall variability on run-off modelling, *Journal of Hydrology*, Vol. 173, pp. 309-326

Foresti, L. & Pozdnoukhov, A. (2011). Exploration of alpine orographic precipitation patterns with radar image processing and clustering techniques, *Meteorological Applications*, Vol. 18, No. 4, DOI:10.1002/met.272

Fox, N. I. and Wikle, C. K. (2005). A Bayesian quantitative precipitation nowcasting scheme, *Weather and Forecasting*, Vol. 20, pp. 264-275

French, M. N. & Krajewski, W. F. (1994) A model for real-time quantitative rainfall forecasting using remote sensing: Part 1 Formulation, Water Resources Research, Vol. 30, No. 4, pp. 1075-1083

Gebremichael, M. & Krajewski, W. (2004). Assessment of the statistical characterisation of small-scale rainfall variability from radar: analysis of TRMM ground validation datasets, *Journal of Applied Meteorology*, Vol. 43, pp. 1180 – 1199

Georgakakos, K. P. & Bras, R. L. (1984a). A hydrologically useful station precipitation model: 1. Formulation, *Water Resources Research*, Vol. 20, No. 11, pp. 1585-1596

Georgakakos, K. P. & Bras, R. L. (1984b). A hydrologically useful station precipitation model: 2. Case studies, *Water Resources Research*, Vol. 20, No. 11, pp. 1597-1610

Georgakakos, K. P. & Kavvas, M. L. (1987). Precipitation analysis, modelling and prediction in hydrology, Reviews of Geophysics, Vol. 25, No. 2, pp. 163-178

Germann, U., Berenguer, M., Sempere-Torres, D. & Zappa, M. (2009). REAL —ensemble radar precipitation estimation for hydrology in a mountainous region, *Quarterly Journal of the Royal Meteorological Society*, Vol. 135, No. 639, pp. 445–456

Germann, U. & Zawadzki, I. (2002). Scale-dependence of the predictability of precipitation from continental radar images. Part 1: Description of the methodology, *Monthly Weather Review*, Vol. 130, pp. 2859–2873

Germann, U. & Zawadzki, I. (2004). Scale dependence of the predictability of precipitation from continental radar images. Part II: Probability forecasts, *Journal of Applied Meteorology*, Vol. 43, pp. 74–89

Germann, U., Zawadzki, I. & Turner, B. (2006). Predictability of precipitation from continental radar images. Part IV: Limits to prediction, *Journal of Atmospheric Science*, Vol. 63, pp. 2092–2108

Grecu, M. & Krajewski, W. F. (2000). A large-sample investigation of statistical procedures for radar-based short-term quantitative precipitation forecasting *Journal of Hydrology*, Vol. 239, pp. 69–84

Golding, B.W. (1998). Nimrod: a system for generating automated very short range forecasts, *Meteorological Applications*, Vol. 5, pp. 1-16

Golding, B. (2009). Long lead time flood warnings: reality or fantasy? *Meteorological Applications*, Vol. 16, pp. 3-12

Gustafsson, N., Berre, L., Hornquist, S., Huang, X. -Y., Lindskog, M., Navascues, B., Mogensen, K. S. and Thorsteinsson, S. (2001). Three-dimensional variational data assimilation for a limited area model. Part I: General formulation and the background error constraint, *Tellus A*, Vol. 53, pp. 425-446

Haggett, C. M. (1986). The use of weather radar for flood forecasting in London, *Proceedings of Conference of River Engineers 1986*, Cranfield, 15th-17th July 1986, Ministry of Agriculture, Fisheries and Food, London

Han, L., Fu, S., Zhao, L., Zheng, Y., Wang, H. & Lin, Y. (2009). 3D convective storm identification, tracking, and forecasting — An enhanced TITAN algorithm, *Journal of Atmospheric and Oceanic Technology*, Vol. 26, pp. 719–732

Handwerker, J. (2002). Cell tracking with TRACE3D - a new algorithm, *Atmospheric Research*, Vol. 61, pp. 15–34

Harris, D., Foufoula-Georgiou, E., Droegemeier, K. K. & Levit, J. J. (2001). Multiscale statistical properties of a high-resolution precipitation forecast, *Journal of Hydrometeorology*, Vol. 2, pp. 406–418

Hilst, G. R. & Russo, J. A. (1960). An objective extrapolation technique for semi-conservative fields with an application to radar patterns, *Tech. Memo 3*, The Travelers Research Center, Hartford, CT, 34 pp

Hitschfeld, W. and Bordan, J. (1954). Errors inherent in the radar measurement of rainfall at attenuating wavelengths, *Journal of Meteorology*, Vol. 11, pp. 58-67

Hollingsworth A. & Lonnberg P. (1986). The statistical structure of short-range forecast errors as determined from radiosonde data. 1. The wind-field. *Tellus A*, Vol. 38, pp. 111-136

Honda, Y., Nishijima, M., Koizumi, K., Ohta, Y., Tamiya, K., Kawabata, T. & Tsuyuki, T. (2005). A pre-operational variational data assimilation system for a non-hydrostatic model at the Japan Meteorological Agency: Formulation and preliminary results, *Quarterly Journal of the Royal Meteorological Society*, Vol. 131, pp. 3465-3475

Horn, B. K. P. & Schunck, B. G. (1981). Determining optical flow, *Artificial Intelligence*, Vol. 17, pp. 185-203

Huang, X., Xiao, Y. Q., Barker, D. M., Zhang, X., Michalakes, J., Huang, W., Henderson, T., Bray, J., Chen, Y., Ma, Z., Dudhia, J., Guo, Y. R., Zhang, X., Won, D. J., Lin, H. C. & Kuo, Y. H. (2009). Four-dimensional variational data assimilation for WRF: Formulation and preliminary results. *Monthly Weather Review*, Vol. 137, pp. 299-314

Hudlow, M. D., Farnsworth, R. K. & Green, D. R. (1981). Hydrological forecasting requirements for precipitation data from space measurements, In: *Precipitation Measurement from Space, Workshop Report*, eds D. Atlas and O. W. Thiele, Goddard Space Flight Center, National Aeronautics and Space Administration, pp. D23-D30

Hunter, I. M. (1954). Polarization of radar echoes from meteorological precipitation, *Nature*, Vol. 173, pp. 165-166

Ivanov, V. Y., Vivoni, E. R., Bras, R. L. & Entekhabi, D. (2004). Preserving high-resolution surface and rainfall data in operationals-scale basin hydrology: A fully-distributed physically-based approach, *Journal of Hydrology*, Vol. 298, pp. 80-111

Johnson, J. T., MacKeen, P. L., Witt, A., Mitchell, E. D., Stumpf, G. J., Eilts, M. D., & Thomas, K. W. (1998). The Storm Cell Identification and Tracking algorithm: An enhanced WSR-88D algorithm, *Weather and Forecasting*, Vol. 13, pp. 263-276

Jones, C. D. & Macpherson, B. (1997). A latent heat nudging scheme for the assimilation of precipitation data into an operational mesoscale model, *Meteorological Applications*, Vol. 4, pp. 269-277

Jordan, P. W., Seed, A. W. & Weinmann, P. E. (2003). A stochastic model of radar measurement errors in rainfall accumulations at catchment scale, *Journal of Hydrometeorology*, Vol. 4, pp. 841 – 855

Kessler, E. (1966). Computer program for calculating average lengths of weather radar echoes and pattern bandedness, *Journal of Atmospheric Science*, Vol. 23, pp. 569-574

Kessler, E. & Russo, J. A. (1963). Statistical properties of weather radar echoes, *Proceedings 10th Weather Radar Conference*, pp. 25-33, Washington, D.C., American Meteorological Society, Boston, Massachusetts

Kober, K., Craig, G. C., Keil, C. & Dörnbrack, A. (2011). Blending a probabilistic nowcasting method with a high-resolution numerical weather prediction ensemble for convective precipitation forecasts, *Quarterly Journal of the Royal Meteorological Society*, DOI:10.1002/qj.939

Kong, F., Xue, M., Thomas, K. W., Wang, Y., Brewster, K. A., Wang, X., Gao, J., Weiss, S. J., Clark, A. J., Kain, J. S., Coniglio, M. C. & Du, J. (2011). CAPS multi-model storm-scale ensemble forecast for the NOAA HWT 2010 spring experiment, *24th Conference on Weather and Forecasting, 20th Conference on Numerical Weather Prediction*, paper 457, American Meteorological Society

Krajewski, W. F. & Georgakakos, K. P. (1985). Synthesis of radar rainfall data, *Water Resources Research*, Vol. 21, No. 5, pp. 764-768

Krajewski, W. F., Venkataraman, L., Georgakakos, K. P. & Jain, S. C. (1991). A Monte Carlo study of rainfall sampling effect on a distributed catchment model, *Water Resources Research*, Vol. 27, pp. 119-128

Krzysztofowicz, R. (1983). Why should a forecaster and a decision maker use Bayes theorem?, *Water Resources Research*, Vol. 19, No. 2, pp. 327-336

Krzysztofowicz, R. (1993). A theory of flood warning systems, *Water Resources Research*, Vol. 29, No. 12, pp. 3981-3994

Krzysztofowicz, R. (1998). Probabilistic hydrometeorological forecasts: Toward a new era in operational forecasting, *Bulletin of the American Meteorological Society*, Vol. 79, No. 2, pp. 243-51

Krzysztofowicz, R. (1999). Bayesian theory of probabilistic forecasting via deterministic hydrologic model, *Water Resources Research*, Vol. 35, No. 9, pp. 2739-2750

Krzysztofowicz, R. (2001). The case for probabilistic forecasting in hydrology, *Journal of Hydrology*, Vol. 249, pp. 2–9

Laroche, S. & Zawadzki, I. (1995). Retrievals of horizontal winds from single-Doppler clear-air data by methods of cross correlation and variational analysis, *Journal of Atmosheric and Oceanic Technology*, Vol. 12, pp. 721–738

Langille, R.C. & Gunn, K. L. S. (1948). Quantitative analysis of vertical structure in precipitation, *Journal of Meteorology*, Vol. 5, pp. 301-304

Lean, H. W., Clark, P. A., Dixon, M., Roberts, N. M., Fitch, A., Forbes, R. & Halliwell, C. (2008). Characteristics of high-resolution versions of the Met Office unified model for forecasting convection over the United Kingdom, *Monthly Weather Review*, Vol. 136, pp. 3408-3424

Lee, G. W. (2006). Sources of errors in rainfall measurements by polarimetric radar: Variability of drop size distributions, observational noise, and variation of relationships between R and polarimetric parameters. *Journal of Atmospheric and Oceanic Technology*, Vol. 23, pp. 1005-1028

Lee, G. W. & Zawadzki, I. (2005a): Variability of drop size distributions: Time scale dependence of the variability and its effects on rain estimation. *Journal of Applied Meteorology*, Vol. 44, pp. 241-255

Lee, G. W. & Zawadzki, I. (2005b). Variability of drop size distributions: Noise and noise filtering in disdrometric data, *Journal of Applied Meteorology*, Vol. 44, pp. 634-652

Lee, G. W. & Zawadzki, I. (2006). Radar calibration by gage, disdrometer and polarimetry: Theoretical limit caused by the variability of drop size distribution and application to fast scanning operational radar data. *Journal of Hydrology*, Vol. 328, pp. 83-97

Lee, G., Seed, A. W. & Zawadzki, I. (2007). Modeling the variability of drop size distributions in space and time, *Journal of Applied Meteorology and Climatology*, Vol. 46, pp. 742-756

Lee, T. & Georgakakos, K. P. (1990). A two-dimensional stochastic-dynamical quantitative precipitation forecasting model, *Journal of Geophysical Research*, Vol. 95, pp. 2113-2126

Lee, H. C., Bellon, A., Kilambi, A. & Zawadzki, I. (2009). McGill Algorithm for Precipitation nowcasting by Lagrangian Extrapolation (MAPLE) applied to the South Korean radar network. Part 1: Sensitivity studies of the Variational Echo Tracking (VET) technique, *Proceedings of the 34th Radar Conference*, American Meteorological Society

Lewin, J. (1986). The control of spillway gates during floods, *Water Services*, Vol. 90, No. 1081, pp. 93-95

Li, L., Schmid, W. & Joss, J. (1995). Nowcasting of motion and growth of precipitation with radar over a complex orography, *Journal of Applied Meteorology*, Vol. 34, No. 6, pp. 1286-1300

Li, P. W., Wong, W. K., Chan, K.Y., & Lai, E. S. T. (2000). SWIRLS – an evolving nowcasting system, *Hong Kong Observatory Technical Note No. 100*, Hong Kong Observatory, 134A Nathan Road, Kowloon, Hong Kong

Liang, Q., Feng, Y., Deng, W., Hu, S., Huang, Y., Zeng, Q. & Chen, Z. (2010). A composite approach of radar echo extrapolation based on TREC vectors in combination with model-predicted winds, *Advances in Atmospheric Sciences*, Vol. 27, No. 5, pp. 1119-1130

Ligda, M. G. H. (1951). Radar storm observation, In: *Compendium of Meteorology*, T. F. Malone, (Ed.), 1265-1282, American Meteorological Society

Ligda, M. G. H. (1953). The horizontal motion of small precipitation areas as observed by radar, *Tech. Rep. 21*, Department of Meteorology, M.I.T., 60 pp., Available from Library, Massachusetts Institute of Technology, 77 Massachusetts Ave., Cambridge,MA 02139

Lin, Y., Ray, P. & Johnson, K. (1993). Initialization of a modelled convective storm using Doppler radar derived fields, *Monthly Weather Review*, Vol. 121, pp. 2757-2775

Llort, X., Velasco-Forero, C., Roca-Sancho, J. & Sempere-Torres, D. (2008) Characterization of uncertainty in radar-based precipitation estimates and ensemble generation, *Proceedings of the Fifth European conference on radar in meteorology and hydrology* (ERAD 2008)

Lorenc, A. C. (1981). A global three-dimensional multivariate statistical interpolation scheme, *Monthly Weather Review*, Vol. 109, pp. 701-721

Lorenc, A. C., Ballard, S. P., Bell, R. S., Ingleby, N. B., Andrews, P. L. F., Barker, D. M., Bray, J. R., Clayton, A. M., Dalby, T., Li, D., Payne, T. J., & Saunders, F. W. (2000). The Met. Office global 3-Dimensional variational data assimilation scheme, *Quarterly Journal of the Royal Meteorological Society*, Vol. 126, pp. 2991-3012

Lorenz, E. N. (1963). Deterministic nonperiodic flow, *Journal of Atmospheric Sciences*, Vol. 20, pp. 120-141

Lorenz, E. N. (1973). On the existence of extended range predictability, *Journal of Applied Meteorology*, Vol. 12, pp. 543-546

Lovejoy, S. & Schertzer, D. (1985). Generalized Scale Invariance in the Atmosphere and Fractal Models of Rain, *Water Resources Research*, Vol. 21, pp. 1233-1250

Lovejoy, S. & Schertzer, D. (1986). Scale Invariance, Symmetries, Fractals and Stochastic Simulation of Atmospheric Phenomena, *Bulletin of the American Meteorological Society*, Vol. 67, pp. 21-32

Macpherson, B., Wright, B. J., Hand, W. H., & Maycock, A. J. (1996). The impact of MOPS moisture data in the UK Meteorological Office mesoscale data assimilation scheme, *Monthly Weather Review*, Vol. 124, No. 8, pp. 1746-1766

Marsan, D., Schertzer, D. & Lovejoy, S. (1996). Causal space-time multifractal processes: predictability and forecasting rain fields, *Journal of Geophysical Research*, 101, 21D, pp. 26333-26346

Marshall, J. S. & Palmer, W. McK. (1948). The distribution of rain-drops with size, *Journal of Meteorology*, Vol. 5, pp. 165-166

Mecklenburg, S., Joss, J. & Schmid, W. (2000). Improving the nowcasting of precipitation in an Alpine region with an enhanced radar echo tracking algorithm, *Journal of Hydrology*, Vol. 239, pp. 46-68

Mecklenburg, S., Bell, V. A., Carrington, D. S., Cooper, A. M., Moore, R. J., & Pierce, C. E. (2001). Applying COTREC-derived rainfall forecasts to the rainfall-runoff model PDM—estimating error sources, *Proceedings of the 30th International Conference on Radar Meteorology*, Munich, Germany, 19-24 July 2001, American Meteorological Society

Megenhardt, D. L., Mueller, C., Trier, S., Ahijevych, D. & Rehak, N. (2004). NCWF-2 Probabilistic Forecasts, Preprints, 11th Conference on Aviation, Range, and Aerospace Meteorology, 5.2, American Meteorological Society, Available online at http://ams.confex.com/ams/pdfpapers/81993.pdf

Mellor, D., Sheffield, J., O'Connell, P. E. & Metcalfe, A. V. (2000a). A stochastic space–time rainfall forecasting system for real time flow forecasting, I: development of MTB conditional rainfall scenario generator, *Hydrology and Earth Systems Science*, Vol. 4, No. 4, pp. 603–615

Mellor, D., Sheffield, J., O'Connell, P.E. & Metcalfe, A.V. (2000b). A stochastic space–time rainfall forecasting system for real time flow forecasting, II: application of SHETRAN and ARNO rainfall runoff models to the Brue catchment, *Hydrology and Earth Systems Science*, Vol. 4, No. 4, pp. 617–626

Mittermaier, M. P. (2007). Improving short-range high-resolution model precipitation forecast skill using time-lagged ensembles, *Quarterly Journal of the Royal Meteorological Society*, Vol. 133, No. 627, pp. 1487–1500

Montmerle, T. & Faccani, C. (2009). Mesoscale assimilation of radial velocities from Doppler radars in a pre-operational framework, *Monthly Weather Review*, Vol. 137, pp. 1939-1953

Moore, R. J. (1987). Towards more effective use of radar data for flood forecasting, In: *Weather Radar and Flood Forecasting*, V. K. Collinge and C. Kirby (Eds.), 223-238, John Wiley & Sons Ltd., Chichester

Moore, R. J., Bell, V. A., Cole, S. J. & Jones, D. A. (2007). Rainfall-runoff and other modelling for ungauged/low-benefit locations. *Science Report – SC030227/SR1*, Research Contractor: CEH Wallingford, Environment Agency, Bristol, UK, 249pp.

Morcrette, C., Lean, H., Browning, K., Nicol, J., Roberts, N., Clark, P., Russell, A. & Blyth, A. (2007). Combination of mesoscale and synoptic mechanisms for triggering an isolated thunderstorm: Observational case study of CSIPIOP 1, *Monthly Weather Review*, Vol. 135, pp. 3728-3749, DOI: 10.1175/2007MWR2067.1

Murphy, A. H. & Carter, G. M. (1980). On the comparative evaluation of objective and subjective precipitation probability forecasts in terms of economic value. *Preprints, Eighth Conference on Weather Forecasting and Analysis*, pp. 478-487, American Meteorological Society, Denver, Colorado, 10-13 June

Newell, R. E., Geotis, S. G., Stone, M. L. & Fleisher, A. (1955). How round are raindrops? *Proceedings of the Fifth Conference on Radar Meteorology*, pp. 261-268, American Meteorological Society, Asbury Park, New Jersey

Noel, T. M. & Fleisher, A. (1960). *The linear predictability of weather radar signals*, Research Rep. 34, Department of Meteorology, M.I.T., 46 pp., Available from Library, Massachusetts Institute of Technology, 77 Massachusetts Ave., Cambridge, MA 02139

Noonan, G. A. (1987). An operational flood warning system, In: *Weather Radar and Flood Forecasting*, 109-126, V. K. Collinge and C. Kirby (Eds.), Wiley, Chichester

Norman, K., Seed, A. & Pierce, C. (2010) A comparison of two radar rainfall ensemble generators, *Proceedings of the Sixth European Conference on Radar in Meteorology and Hydrology* (ERAD 2010)

Novak, P. (2007). The Czech Hydrometeorological Institute's severe storm nowcasting system, *Atmospheric Research,* Vol. 83, pp. 450-457

Obled, C., Wendling, J. & Bevin, K. (1994). The sensitivity of hydrological models to spatial rainfall patterns: an evaluation using observed data, *Journal of Hydrology*, Vol. 159, pp. 305-333

Ogden, F. L. & Julien, P. Y. (1994). Runoff model sensitivity to radar rainfall resolution, *Journal of Hydrology*, Vol. 158, pp. 1-18

Ott, E., Hunt, B. R., Szunyogh, I., Zimin, A. V., Kostelich, E. J., Corazza, M., Kalnay, E., Patil, D. J. & Yorke, J. A. (2004). A Local ensemble Kalman filter for atmospheric data assimilation, *Tellus A*, Vol. 56, 415-428

Pappenberger, F., Scipal, K. & Buizza, R. (2008). Hydrological aspects of meteorological verification, Atmosperic Science Letters, Vol. 9, No. 2, pp. 43–52

Panziera, L., Germann, U., Gabella, M. & Mandapaka, P.V. (2011). NORA–Nowcasting of Orographic Rainfall by means of Analogues, *Quarterly Journal of the Royal Meteorological Society*, Vol. 137, pp. 2106-2123

Pegram, G. G. S. & Clothier, A. N. (2001). Downscaling rainfields in space and time, using the String of Beads in time series mode, *Hydrology and Earth Systems Science*, Vol. 5, No. 2, pp. 175-186, doi:10.5194/hess-5-175-2001

Pessoa, M. L., Raael, L. B. & Earle, R. W. (1993). Use of weather radar for flood forecasting in the Sieve river basin: A sensitivity analysis, *Journal of Applied Meteorology*, Vol. 32, pp. 462-475

Peura, M. & Hohti, H. (2004). Optical flow in radar images, *Proceedings of the Third European Conference on Radar Meteorology (ERAD)*, pp. 454-458, Visby, Island of Gotland, Sweden, 6-10 September 2004

Pierce, C. E., Norman, K. and Seed, A. (2011). Use of ensemble radar estimates of precipitation rate within a stochastic, quantitative precipitation nowcasting algorithm, *Proceedings of the eighth International Symposium on Weather Radar and Hydrology*, Exeter, UK, April 2011, to appear in IAHS red book

Pierce, C. E., Bowler, N., Seed, A., Jones, A., Jones, D. & Moore, R. J. (2004). Use of a stochastic precipitation nowcast scheme for fluvial flood forecasting and warning. *Proceedings of the Sixth International Symposium on Hydrological Applications of Weather Radar*, 2-4 February 2004, Melbourne, Australia

Poli, V., Alberoni, P.P. & Cesari, D. (2008). Intercomparison of two nowcasting methods: preliminary analysis, *Meteorology and Atmospheric Physics*, Vol. 101, pp. 229-244, doi: 10.1007/s00703-007-0282-3

Rawlins, F. R., Ballard, S. P., Bovis, K. R., Clayton, A. M., Li, D., Inverarity, G. W., Lorenc, A. C., & Payne, T. J. (2007). The Met Office global four-dimensional data assimilation system, *Quarterly Journal of the Royal Meteorological Society*, Vol. 133, pp. 347-362

Rennie, S. J., Dance, S. L., Illingworth, A. J., Ballard, S. P. & Simonin, D. (2010). 3D-Var assimilation of insect-derived Doppler radar radial winds in convective cases using a high-resolution model, *Monthly Weather Review*, Vol. 139, pp. 1148-1163

Rigo, T., Pineda, N. & Bech, J. (2010). Analysis of warm season thunderstorms using an object-oriented tracking method based on radar and total lightning data, Natural Hazards and Earth System Sciences, Vol. 10, pp. 1881-1893

Rinehart, R. E. & Garvey, E. T. (1978). Three-dimensional storm motion detection by conventional weather radar, *Nature*, Vol. 273, pp. 287-289

Rinehart, R. E. (1981). A pattern-recognition technique for use with conventional weather radar to determine internal storm motions, In: *Recent Progress in Radar Meteorology*, R. Carbone (Ed.), 105–118, National Center for Atmospheric Research

Roberts, N. M. & Lean, H. W. (2008). Scale-selective verification of rainfall accumulations from high-resolution forecasts of convective events, *Monthly Weather Review*, Vol. 136, pp. 78-97

Roberts, N. M., Cole, S. J., Forbes, R. M., Moore, R. J. & Boswell, D. (2009). Use of high-resolution NWP rainfall and river flow forecasts for advance warning of the Carlisle flood, north-west England, *Meteorological Applications*, Vol. 16, pp. 23-34

Roca-Sancho, J., Berenguer, M., Zawadzki, I., & Sempere-Torres, D. (2009). Characterization of the error structure of precipitation nowcasts, *Proceedings of the 34th AMS Radar Conference*, American Meteorological Society

Rossa, A. M., Liechti, K., Zappa, M., Bruen, M., Germann, U., Haase, G., Keil, C. & Krahe, P. (2011). COST 731 Action: A review on uncertainty propagation in advanced hydro-meteorological forecast systems, Atmospheric Research, Vol. 100, pp. 150-167

Russo, J. A. & Bowne, N. E. (1962). Linear extrapolation as a meteorological forecast tool when applied to radar and cloud ceiling patterns, *Proceedings of the ninth weather radar conference*, Kansas City, Mo., October 23-26, 1961

Ruzanski, E., Chandrasekar, V., & Wang, Y. (2011). The CASA nowcasting system, *Journal of Atmospheric and Oceanic Technology*, Vol. 28, pp. 640-655

Ryde, J. W. (1946). The attenuation and radar echoes produced at centimetre wave-lengths by various meteorological phenonema, *Meteorological Factors in Radio-Wave Propagation*, Physical Society of London, pp. 169-189

Saito, K., Fujita, T., Yamada, Y., Ishida, J., Kumagai, Y., Aranami, K., Ohmori, S., Nagasawa, R., Kumagai, S., Muroi, C., Kato, T., Eito, H. & Yamazaki, Y. (2006). The operational JMA nonhydrostatic mesoscale model, *Monthly Weather Review*, Vol. 134, pp. 1266-1298

Schaake, J. C. & Larson, L. (1998). Ensemble streamflow prediction (ESP): Progress and research needs, *Special Symposium on Hydrology*, J1.3, 410-413, American Meteorological Society, Phoeniz, Arizona

Schellart, A. N. A., Rico-Ramirez, M. A., Liguori, S. & Saul, A.J. (2009). Quantitative precipitation forecasting for a small urban area: use of radar nowcasting, *8th International Workshop on precipitation in urban areas, Rainfall in the urban context: Forecasting, Risk and Climate Change*, 10-13 December, 2009, St. Moritz, Switzerland

Schellekens, J., Minett, A. R. J., Reggiani, P., Weerts A. H., Moore, R. J., Cole, S. J., Robson, A. J., Bell, V. A. (2010). *Hydrological modelling using convective scale rainfall modelling – phase 3*. Project: SC060087/R3, Authors. Research Contractor: Deltares and CEH Wallingford, Environment Agency, Bristol, UK, 231pp. http://publications.environment-agency.gov.uk/pdf/SCHO0210BRYT-e-e.pdf

Schenkman, A., Xue, M., Shapiro, A., Brewster, K. & Gao, J. (2011a). The analysis and prediction of the 8-9 May 2007 Oklahoma tornadic mesoscale convective system by assimilating WSR-88D and CASA radar data using 3DVAR, *Monthly Weather Review*, Vol. 139, pp. 224-246

Schenkman, A., Xue, M., Shapiro, A., Brewster, K. & Gao, J. (2011b). Impact of CASA radar and Oklahoma mesonet data assimilation on the analysis and prediction of tornadic mesovortices in a MCS, *Monthly Weather Review*, Vol. 139, pp. 3422-3445

Schertzer, D., Lovejoy, S., Schmitt, F., Chigirinskaya, Y. & Marsan, D. (1997). Multifractal cascade dynamics and turbulent intermittency, *Fractals*, Vol. 5, No. 3, pp. 427-471

Schmid, W., Mecklenburg, S. & Joss, J. (2000). Short-term risk forecasts of severe weather, *Physics and Chemistry of the Earth, Part B*, Vol. 25, pp. 1335-1338

Seed, A. W., Srikanthan, R. & Menabde, M. (1999). A space and time model for design storm rainfall, *Journal of Geophysical Research*, Vol. 104, No. D24, pp. 31623-31630

Seed, A. W. (2003). A dynamic and spatial scaling approach to advection forecasting, *Journal of Applied Meteorology*, Vol. 42, pp. 381-388

Seity, Y., Brousseau, P., Malardel, S., Hello, G., Benard, P., Boutiier, F., Lac, C. & Mason, V. (2011). The AROME-France convective scale operational model, *Monthly Weather Review*, Vol. 139, pp. 976-991

Seo, D.-J., Perica, S., Welles, E., & Schaake, J. C. (2000). Simulation of precipitation fields from probabilistic quantitative precipitation forecast, *Journal of Hydrology*, Vol. 239, No. 1, pp. 203-229

Shah, S. M. S., O'Connell, P. E., & Hosking, J. R. M. (1996). Modelling the effects of spatial variability in rainfall on catchment response. II. Experiments with distributed and lumped models, *Journal of Hydrology*, Vol. 175, pp. 89-111

Sharif, H. O., Yates, D., Roberts, R. & Mueller, C. (2006). The use of an automated nowcasting system to forecast flash floods in an urban watershed, *Journal of Hydrometeorology*, Vol. 7, pp. 190–202

Smith, T. L., Benjamin, S. G., Brown, J. M., Weygandt, S. S., Smirnova, T. & Schwartz, B. E. (2008). Convection forecasts from the hourly updated, 3-km high resolution Rapid Refresh Model, *Preprints, 24th Conference on severe local storms*, Savannah, GA, American Meteorological Society, Available online at http://ams.confex.com/ams/24SLS/techprogram/paper_142055.htm

Snook, N., Xue, M. & Jung, J. (2011). Analysis of a tornadic meoscale convective vortex based on ensemble Kalman filter assimilation of CASA X-band and WSR-88D radar data, *Monthly Weather Review*, Vol. 139, pp. 3446-3468

Snook, N., Xue, M. & Jung, J. (2012). Ensemble probabilistic forecasts of a tornadic mesoscale convective system from ensemble Kalman filter analyses using WSR-88D and CASA radar data, to appear in *Monthly Weather Review*

Stensrud, D. J., Xue, M., Wicker, L. J., Kelleher, K. E., Foster, M. P., Schaefer, J. T., Schneider, R. S., Benjamin, S. G., Weygandt, S. S., Ferree, J. T. & Tuell, J. P. (2009). Convective-scale warn-on-forecast system. A vision for 2020, *Bulletin of the American Meterological Society*, Vol. 90, pp. 1487-1499

Stephan, K., Klink, S. & Schraff, C. (2008). Assimilation of radar-derived rain rates into the convective-scale model COSMO-DE at DWD, *Quarterly Journal of the Royal Meteorological Society*, Vol. 134, pp. 1315-1326

Stout, G. E. & Neill, J. C. (1953). Utility of radar in measuring areal rainfall, *Bulletin of the American Meteorological Society*, Vol. 34, No. 1, pp. 21-27

Sun, J. (2005a). Initialization and numerical forecasting of a supercell storm observed during STEPS, *Monthly Weather Review*, Vol. 133, pp. 793-164

Sun, J. (2005b). Convective-scale assimilation of radar data: progress and challenges, *Quarterly Journal of the Royal Meteorological Society*, Vol. 31, pp. 3439-3463

Sun, J., Chen, M. & Wang, Y. (2010). A frequent-updating analysis system based on radar, surface, and mesoscale model data for the Beijing 2008 Forecast Demonstration Project, *Weather and Forecasting*, Vol. 25, pp. 1715-1735

Sun, J. & Crook, A. (1994). Wind and thermodynamic retrievals from single-Doppler measurements of a gust front observed during Phoenix II, *Monthly Weather Review*, Vol. 122, pp. 1075-1091

Sun, J. & Crook, N. A. (1997). Dynamical and microphysical retrieval from Doppler radar observations using a cloud model and its adjoint: I. Model development and simulated data experiments, *Journal of Atmospheric Science*, Vol. 54, pp. 1642-1661

Sun, J. & Crook, N. A. (1998). Dynamical and microphysical retrieval from Doppler radar observations using a cloud model and its adjoint: II. Retrieval experiments of an observed Florida convective storm, *Journal of Atmospheric Science*, Vol. 55, pp. 835-852

Sun, J. & Crook, N. A. (2001). Real-time low-level wind and temperature analysis using single WSR-88D data, *Weather Forecasting*, Vol. 16, pp. 117-132

Sun, J., Flicker, D. W. & Lilly, D. K. (1991). Recovery of three-dimensional wind and temperature fields from single-Doppler radar data, *Journal of Atmospheric Science*, Vol. 48, pp. 876-890

Sun, J. & Zhang, Y. (2008). Analysis and prediction of a squall line observed during IHOP using multiple WSR-88D observations, *Monthly Weather Review*, Vol. 136, pp. 2364-2388

Sun, J., Xiao, Q., Trier, S., Weisman, M., Wang, H., Ying, Z., Xu, M. & Zhang, Y. (2012). Sensitivity of 0-12 hour warm-season precipitation forecast over the central United States to model initialization and parameterizations, submitted to *Weather and Forecasting*

Tsonis, A. A. & Austin, G. L. (1981). An evaluation of extrapolation techniques for the short-term prediction of rain amounts, *Atmosphere–Ocean*, Vol. 19, pp. 54–65

Tsonis, A. A. (1989). Chaos and unpredictability of weather, *Weather*, Vol. 44, No. 6, pp. 258-263

Twomey, S. (1953). On the measurement of precipitation by radar, *Journal of Meteorology*, Vol. 10, pp. 601-620

Turner, B. J., Zawadzki, I. & Germann, U. (2004). Predictability of precipitation from continental radar images. Part III: Operational nowcasting implementation (MAPLE), *Journal of Applied Meteorology*, Vol. 43, pp. 231-248

Velasco-Forero, C. A., Sempere-Torres, D., Cassiraga, E. F. & Gómez-Hernández, J. J. (2009). A non-parametric automatic blending methodology to estimate rainfall fields from rain gauge and radar data. *Advances in Water Resources*, Vol. 32, pp. 986-1002

Vivoni, E. R., Ivanov, V. Y., Bras, R. L. & Entekhabi, D. (2005). On the effects of triangulated terrain resolution on distributed hydrologic model response, *Hydrological Processes*, Vol. 19, pp. 2101-2122

Vivoni, E. R., Entekhabi, D., Bras, R. L., Ivanov, V. Y. & Van Horne, M. P. (2006). Extending the predictiability of hydrometeorological flood events using radar rainfall nowcasting, *Journal of Hydrometeorology*, Vol. 7, pp. 660-677

Vivoni, E. R., Entekhabi, D. & Hoffman, R. N. (2007). Error propagation of radar rainfall nowcasting fields through a fully distributed flood forecasting model, *Journal of Applied Meteorology And Climatology*, Vol. 46, pp. 932-940

Wang, J. and contributing authors (2009). *Overview of the Beijing 2008 Olympics project. Part I; Forecast Demonstration Project*, A report to the WMO World Weather Research Programme

Werner, M. & Cranston, M. (2009). Understanding the value of radar rainfall nowcasts in flood forecasting and warning in flashy catchments, *Meteorological Applications*, Vol. 16, No. 1, pp. 41–55

Werner, M., Cranston, M., Harrison, T., Whitfield, D. & Schellekens, J. (2009). Recent developments in operational flood forecasting in England, Wales and Scotland, *Meteorological Applications*, Vol. 16, No. 1, pp. 13–22

Wexler, R. (1951). Theory and observation of radar storm detection, *Compendium of Meteorology*, T. F. Malone, (Ed.), 1283-1289, American Meteorological Society

Wexler, R. & Swingle, D. M. (1947). Radar storm detection, *Bulletin of the American Meteorological Society*, Vol. 28, pp. 159-167

Weygandt, S. S., Benjamin, S. G., Smirnova, T. G. & Brown, J. M. (2008). Assimilation of radar reflectivity data using a diabatic digital filter within the Rapid Update Cycle., *Preprints, 12th Conference on IOAS-AOLS*, 8.4., New Orleans, LA, Amercan Meteorological Society, Available online at http://ams.confex.com/ams/88Annual/techprogram/paper_134081.htm

Winchell, W., Gupta, H. V. & Sorooshian, S. (1998). On the simulation of infiltration- and saturation-excess runoff using radar-based rainfall estimates: Effects of algorithm uncertainty and pixel aggregation, *Water Resources Research*, Vol. 34, pp. 2655-2670

Wilk, K. E. & Gray, K. C. (1970). Processing and analysis techniques used with the NSSL weather radar system, *Preprints, 14th Conference on Radar Meteorology*, pp. 369–374, Tucson, AZ, American Meteorological Society

Wilson, J. W. (1966). Movement and predictability of radar echoes, *Tech. Memo ERTM-NSSL-28*, National Severe Storms Laboratory, 30 pp., Available from National Information Service, Operations Division, Springfield, VA 22161

Wilson, C. B., Valdes, J. B., & Rodriguez-Iturbe, I. (1979). On the influence of the spatial distribution of rainfall on storm runoff, Water Resources Research, Vol. 15, No. 2, pp. 321-328

Wilson, J. W., Crook, N. A., Mueller, C. K., Sun, J. & Dixon, M. (1998). Nowcasting thunderstorms : a status report, *Bulletin of the American Meteorological Society*, Vol. 79, pp. 2079-2099

Wilson, J. W., Feng, Y., Chen, M., & Roberts, R. (2010). Nowcasting challenges during the Beijing Olympics; Successes, failures, and implications for future nowcasting systems, *Weather Forecasting*, Vol. 25, pp. 1691-1714

Witt, A. & Johnson, J. T. (1993). An enhanced storm cell identification and tracking algorithm, *Preprints, 26th International Conference on Radar Meteorology*, pp. 514–521, Norman, OK, American Meteorological Society

Xiao, Q., Kuo, Y. -H., Sun, J., Lee, W. -C., Lim, E., Guo, Y., & Barker, D. M. (2005). Assimilation of Doppler radar observations with a regional 3D-VAR system: Impact of Doppler velocities on forecasts of a heavy rainfall case, *Journal of Applied Meteorology*, Vol. 44, pp. 768-788

Xiao, Q., Lim, E., Won D. -J, Sun, J., Lee, W. -C., Lee, M. -S., Lee, W. -J., Cho, J. -Y., Kuo, Y. -H., Barker, D. M., Lee, D. -K., Lee, H. -S. (2008). Doppler radar data assimilation in KMA's operational forecasting, *Bulletin of the American Meteorological Society*, Vol. 89, pp. 39-43

Xue, M., Kong, F., Thomas, K. W., Wang, Y., Brewster, K., Gao, J., Wang, X., Weiss, S. J., Clark, A. J., Kain, J. S., Coniglio, M. C., Du, J., Jensen, T. L. & Kuo, Y. H. (2011). CAPS Real time storm scale ensemble and high resolution forecasts for the NOAA hazardous weather testbed 2010 spring experiment, *24th Conference on Weather and Forecasting, 20th Conference on Numerical Weather Prediction*, paper 9A.2, American Meteorological Society

Yeung, H.- Y, Man, C., Chan, S. -T. & Seed, A. W. (2011). Application of radar-raingauge co-Kriging to improve QPE and quality control of real-time rainfall data, *Proceedings of the eighth International Symposium on Weather Radar and Hydrology*, Exeter, UK, April 2011, to appear in IAHS red book

Zappa, M., Rotach, M. W., Arpagaus, M., Dorninger, M., Hegg, C., Montani, A., Ranzi, R., Ament, F., Germann, U., Grossi, G., Jaun, S., Rossa, A., Vogt, S., Walser, A., Wehrhan, J. & Wunram, C. (2008). MAP D-PHASE: real-time demonstration of hydrological ensemble prediction systems, *Atmos. Science Letters*, Vol. 9, No. 2, pp. 80-87

Zawadzki, I. (1973). Statistical properties of precipitation patterns, *Journal of Applied Meteorology*, Vol. 12, pp. 459–472

Zawadzki, I., Morneau, J. & Laprise, R. (1994). Predictability of precipitation patterns: an operational approach, *Journal of Applied Meteorology*, Vol. 33, pp. 1562 – 1571

Zittel, W. D. (1976). Computer applications and techniques for storm tracking and warning, *Preprints, 17th Conference on Radar Meteorology*, pp. 514–521, Seattle, WA, American Meteorological Society

5

Measuring Snow with Weather Radar

Elena Saltikoff
Finnish Meteorological Institute
Finland

1. Introduction

People of warm climate tend to think of snow as something rather rare and exotic. However, most weather radars operating at mid-latitudes measure snow every day, at some altitude. Even at relatively low elevation angles the edges of a PPI image are often measured above the freezing level. Fig. 1 shows radar measurements from a night when surface minimum temperature was +13 °C – still, the majority of radar measurement volume was filled with snow.

In many meteorological classifications hydrometeors are divided in two or three classes (rain, wet snow and dry snow, see Fig. 1.). On the other hand, we have been told that there are not two identical snowflakes. Between these extremes are the snowflake type classifications such as those by Ukichiro Nakaya (Nakaya, 1954). From his work we can learn how, based on temperature and humidity, snow crystals can take the shape of needles, columns, plates, stars, rosettes and dendrites, only to name a few. They can also join each other in a process called aggregation, and they can be covered in icing in a process called riming.

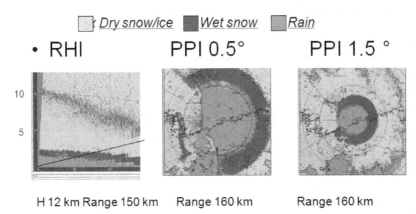

Fig. 1. Hydrometeor classification 30 August 2009 00:45 UTC in Vantaa, Finland. RHI to 150 km in range, 12 km in height (left). PPI to 160 km in range, 0.5 degrees in elevation (middle) and 1.5 degrees in elevation (right). Cyan for dry snow, dark blue for melting snow, light blue for rain.

Measuring snowfall with short wavelengths can bring us to the edge of assumption of the radar equation for Rayleigh scattering: are the particles much smaller than the radar wavelength ?

In this chapter, snow will be discussed from viewpoint of a radar meteorologist. Many topics are relevant for operational weather service, others more for the researcher. Increasing use of polarimetric radars is bringing new perspectives to measuring snow with radars.

2. Vertical structure of snowfall

With the usual measuring geometry of a scanning weather radar we have to take into account the vertical structure of precipitation. In warm weather, we measure rain near ground, wet snow above it and dry snow on the top, as can be seen in Fig. 1. A typical reflectivity structure is related to the temperature structure so that we have a maximum just below 0 °C isotherm. Above it, in the snowfall area, reflectivity decreases with an even gradient of approximately 7.5 dBZ/km. This decrease is related to four factors:

- at higher altitudes, it is colder and snow crystals are typically smaller in diameter
- at higher altitudes, the absolute humidity is smaller so the mass of snow per cubic kilometer of cloud is smaller there
- crystals fall down while they grow, so older crystals which have had time to grow large are more likely to be located at lower altitudes
- near the cloud top there may be effects of partial beam overshooting

In the precipitation system of a warm front, the two first factors create also horizontal gradients: the leading edge is in colder and drier air.

When the snowflakes melt, the surface gets wet first while the inner parts are   still of dry snow. The partially-melted, wet snowflakes have approximately the size and fallspeed of snowflakes, but the dielectric properties of water surfaces. Hence the radar reflectivity peaks in the melting layer, a phenomenon also known as the bright band. In the hands of an inexperienced user of radar data, this could lead to an overestimation of precipitation intensity. In a modern weather radar service, the overestimation is corrected using knowledge of the vertical profile of reflectivity (Koistinen et al., 2003). Recently, Giangrande et al. (2005) and Boodoo et al. (2010) have shown, that the parameters of dual-polarization radars can be used effectively to follow the temporal and spatial variation of the melting layer height and thickness. This is especially important in cold and temperate climates, where much of precipitation is associated with fronts, because in frontal situations the temperature gradients are sharp.

In Fig. 2 we see RHI and PPI images in a snowstorm in Finland 2 February 2010. Cloud tops are observed between 6 and 8 km, and reflectivity is growing downwards from there. No bright band is observed, as there is no melting. Temperatures in cloud tops are near -35..-40 °C, at ground -5..-7 °C (based on Tallinn and Jokioinen 00 UTC soundings). The effect of vertical gradient is obvious in the RHI image, but the gradient in PPI is related to two factors: the vertical gradient and the horizontal variation of intensity.

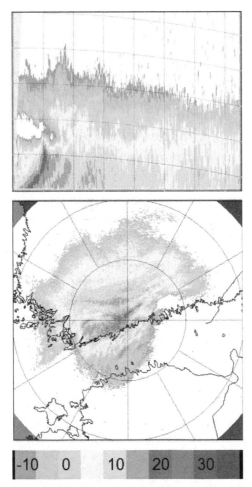

Fig. 2. Radar measurements in a snowstorm in Finland 2 Feb 2010. RHI north of radar to range of 150 km, horizontal lines at every 2 km (upper panel) and PPI with elevation of 1.5 degrees, rings at each 100 km (lower panel). Reflectivity values are between -15 and +25 dBZ.

3. Fallspeeds of solid hydrometeors

Terminal fallspeeds of hydrometeors are influenced by their type and size, and hence Doppler spectra measured with vertically pointing radar can be used for hydrometeor classification. Barthazy and Schefold (2006) showed that the fall velocity of snowflakes consisting of needles or plates is strongly dependent on the riming degree. The average fall velocity of any type of snowflakes of diameter of 1 mm or larger is typically between 1 and 2 m s^{-1}. In cases when hydrometeor types change, or two types of hydrometeors coexist in same measurement volume, Doppler spectra can provide valuable information of cloud physical processes such as riming and aggregation. In addition of academical interest they may provide value in aviation weather services.

4. Clutter cancellation and clear air echoes

One of the main reasons why operational weather services started to use Doppler radars was the use of Doppler signal for clutter cancellation of reflectivity fields. The principle is simple: precipitation has velocity (at least fallspeed and turbulence), while ground clutter does not, and Doppler radar can measure the velocity (or absence thereof). However, when the precipitation is in form of snow, there are some complicating details, and we have to study the filtering process in depth.

A Doppler radar does not measure the true speed of the particles, just the component parallel to the radar beam. When the real wind is nearly perpendicular to the beam (e.g. for northerly winds in east and west of radar), this component is near zero. We can not set the threshold to censor only the bins with exactly zero speed, because even clutter targets have apparent speeds due to different viewing angle of the rotating antenna, trees and masts waving in wind and trucks and lorries in an urban environment. On the other hand, if the threshold of censorship is too high, Doppler filter removes data with near-zero Doppler velocities in areas where wind is perpendicular to beam. Because there is wind shear and measurement is made at different heights at different distances, the direction of missing data is undulating, and hence the gap in reflectivity PPIs is sometimes called *Doppler snake*.

In Fig. 3, it is relatively easy to see the underestimation of reflectivity on a line coiling from west-southwest through the radar and to the opposite side. In this case, there is also a "secondary snake" south of the radar, where the near-zero-velocities are related to folding.

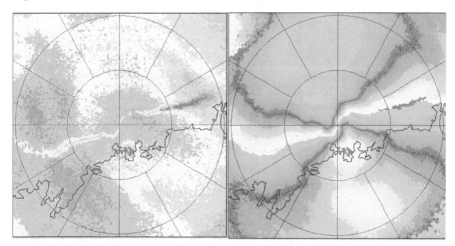

Fig. 3. A Doppler snake case 17 March 2005 03:00 UTC. PPIs of reflectivity (left) and Doppler velocity (right). Wind is from south-southeast, warm colours indicate echoes moving away from the radar and cold colours towards the radar . The velocity field is folded, unambiguous speed 7.6 m/s. A band of weaker reflectivities cutting through the image from southwest to east, at same location as zero Doppler speeds (white). Reflectivity colour scale as in Fig. 2.

It is tempting to try to get rid of the Doppler snake by using a less aggressive Doppler filter. However, amount of residual clutter may increase. Finding the compromise between too aggressive and too weak filter is threading on a fine line, and the selection should be tested in different wind and temperature conditions. Temperature inversions (typical for cloudfree winter days or nights) affect amount of clutter by causing anomalous propagation, which then leads to the increase of ground or sea clutter. Wind affects sea clutter but may also cause blowing snow. Especially the blowing snow falling from trees in hilly areas may give false alarms of ground clutter – it is real snow flying in real wind and hence immune to most clutter cancellation techniques, even though it is not precipitation falling from clouds.

Case of sea clutter is especially annoying. In summer we have nocturnal inversions and anomalous propagation mainly when it is not raining. In winter it is very likely to have on continent cold weather and inversion, and simultaneously rigorous lake effect snow (see section 6.2) over the water areas. This is possible, because the propagation is affected by temperature near the radar, not at the measurement location. Sea clutter is immune to Doppler filtering, but dual polarization measurements can reveal it, as is seen in Fig. 4.

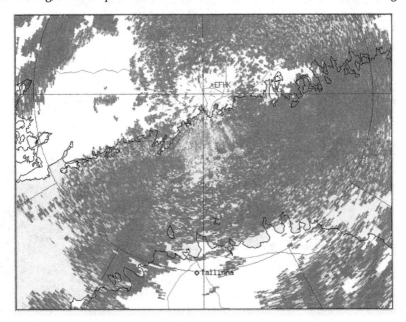

Fig. 4. Sea clutter and lake effect snow seen with polarimetric parameter RhoHV. RhoHV over 0.98 in snow, less than 0.8 in sea clutter. The range ring indicates 100 km from the radar.

5. Dual polarization

While one of the most popular applications of dual-polarisation technology enables one to distinguish the types of the hydrometeors measured (Straka et al., 2000), dual polarization can be used for much more. On the other hand, some applications developed for rain, such as KDP-based algorithms for quantitative precipitation estimates do not work in snow.

When implementing published algorithms to new environments, it is wise to compare the hardware used in the original development work to the platform where it will be implemented. Research community has used S-band radars a lot, while the operational weather services in cold climates use mainly C-band. There may be also difference between simultaneous and alternating transmission of the dual polarization channels.

5.1 Hydrometeor classification

Polarimetric properties of wet snow and single snow crystals are very different from the ones in rain. Hence, these two snow categories are easily distinguishable from rain. However, discrimination between stratiform rain and dry snowflakes (aggregates) is challenging, as relatively low Z and ZDR, and high RhoHV are typical for both (Ryzhkov & Zrnic, 1998). A typical solution in this case is to use the information of height of melting layer, either from sounding, NWP model or radar measurements.

Decision boundaries for dense snow, dry snow, wet snow, rain, dry graupel, wet graupel rain and hail co-existing and hail alone using Z, ZDR, RhoHV, LDR and KDP have been published by e.g. Straka and Zrnic (1993). In many of the parameters, the selected classes overlap, which has encouraged researchers to try fuzzy logic (Bringi & Chandrasekar, 2001). In Fig. 5 we see RHI scans in the same situation as in Fig. 1. The 0 °C isotherm is at 2 km, and a layer of melting snow can be seen below it.

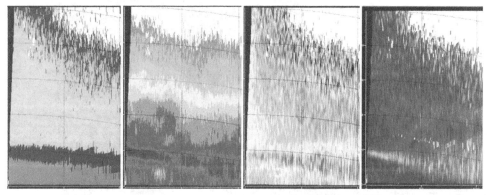

Fig. 5. RHI scans north of Vantaa radar 30 August 2009 00:45 UTC. . Range 100 km, height 10 km, parameters from left to right: Hydrometeor classification, reflectivity Z, differential reflectivity ZDR and copolar correlation factor RhoHV.

5.2 Snow types

In the JPOLE classifier for cold season Ryzhkov et al. (2005) distinguished between Dry aggregated snow: DS and Wet snow: WS, and single crystals. In some cases, this distinction can be made using reflectivity and differential reflectivity alone: wet snow has typically larger reflectivities, and individual crystals larger differential reflectivities. However, the variability of both parameters for both classes is large, and the definitions tend to overlap.

Developing a universal method for snow type classification is even more challenging than finding a representative ZR relation: reliable surface observations of snowflake type are rare, and if they are performed at longer distances from the radar, the snowflakes can change between the radar measurement and the surface observation. Correction for vertical profile of reflectivity is a standard procedure, but correction for vertical profile of snow type is still strongly hypothetical.

6. Characteristic properties of typical snowfall situations

In general, precipitation events can be split to orographic, frontal and convective precipitation. Especially snowfall is often related to warm fronts and lake effect induced convection.

6.1 Warm fronts

Much of snowfall is related to frontal systems of extratropical systems. In their analysis and forecasting, the value of weather radar data lies primarily in the mesoscale structure: detection of the mesoscale bands of heavy snowfall is needed for accurate short term forecasting. The banded structure leads to rapid changes in visibility, and areal differences of accumulated snowfall. Their dynamical structure is complicated, related to negative equivalent potential vorticity (EPV) mainly associated with conditional symmetric instability (CSI), and not always perfectly forecasted by numerical weather prediction models. Hence, identification and extrapolation of movement of these bands using a radar can improve short-range forecasts of extreme events significantly (Nicosia and Grumm, 1999).

Snowfall from warm fronts is also a challenge for a radar meteorologist: forgetting the three-dimensional structure of the frontal system can lead to embarrassing misinterpretation.

In satellite images, we can see the leading edge of frontal system ("warm front shield") and educated meteorologists already know, that arrival of this edge does not mean onset of precipitation. In Fig. 6 the shield at 2 km extends 70 km ahead the surface precipitation. I sincerely hope that everyone using different radar products remembers this, too: the leading edge in products like TOPS, MAX, VIL or even medium-level CAPPI does not indicate the precipitation on ground level. See Fig. 7.

The sloping edge of precipitation area can also be seen in PPIs. In upper panels of Fig. 8, the gap in the centre of the image indicates area where radar beam was below the warm front shield. The gap gets smaller when the surface front approaches the radar. Also, the gap is not a circle but an oval, also indicating the slope of the cloud base.

Warm fronts are ideal for producing Doppler wind profiles (VAD and VVP), because the wind field is usually uniform. In lower right panel of Fig. 8 we have a time series of VVP wind and reflectivity. In this case, the wind shear related to the warm advection is not very strong, and it is hard to distinguish it from wind shear related to friction in the boundary layer. Sharper turning of wind is of interest to aviation weather service. We can also see the shield of overhanging precipitation with approaching front (05-06 UTC) from "+"-signs indicating missing wind barbs. Warm advection can also be seen indirectly: part of the increase of reflectivity at low altitudes around 08 UTC is probably related to temperatures rising to near zero, and snowflakes growing larger.

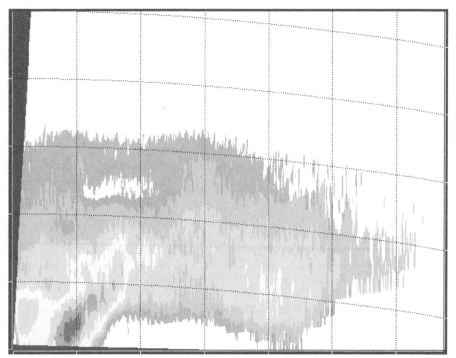

Fig. 6. RHI north of Anjalankoski radar 17 December 2011 12 UTC, warm front approaching from south. Vertical lines at 20 km, horizontal lines at 2 km intervals. Colour scale from -10 dBZ (blue) to 20 dBZ (red) as in Fig. 2.

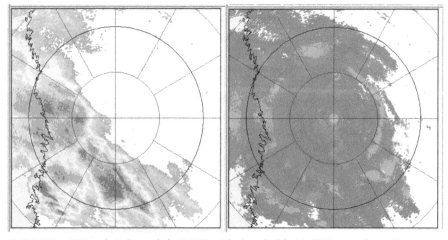

Fig. 7. CAPPI at 500 m height on left, TOPS with threshold -10 dBZ on right. 14 January 2008 05:30 UTC, precipitation on the surface had not yet reached the radar location. Range rings at 50 and 100 km from radar, reflectivity scale as in Fig. 2., green shades for tops in steps of 2 km.

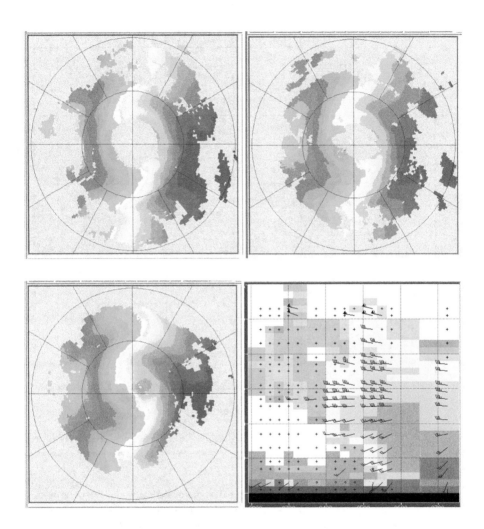

Fig. 8. PPI images of Doppler velocity 30 minutes apart (5:00, 5:30 and 6:00) and time series of wind and reflectivity profiles 02:00 to 07.00 UTC 14 January 2008. Green colours towards radar, reds and yellows away. Reflectivity from -20 dBZ (blue) to +20 dBZ(purple)

In Fig. 9 we see another scale of shear. There is a small wind maximum below 1 km. In this case it lasted for less than 45 minutes. This is an indicator of low level jet related to conveyor belt in the frontal structure. In the real atmosphere, the wind shear related to that is probably larger, as the VVP is averaged over a cylinder of 30 km.

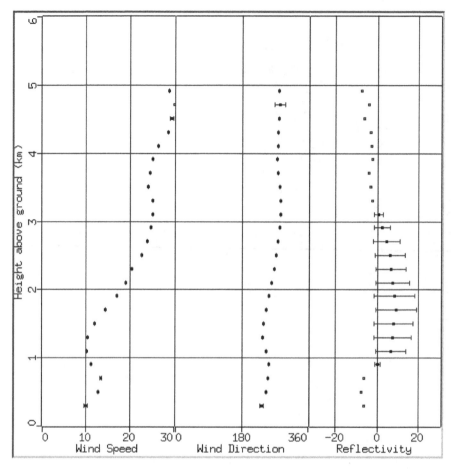

Fig. 9. VVP sounding. Windspeed on left, wind direction in middle and reflectivity on right.

6.2 Lake effect snow

Lake effect snow is a phenomenon observed regularly around open water surfaces in cold weather. The name originated from weather phenomena around the Great Lakes of North America, but it is also observed around bays and straits of sea and great rivers. Lake-effect snowstorms get their energy from the temperature difference between the relatively warm open water and very cold, continental air blowing over the water. These provide the most spectacular outbreaks of boundary layer convection in winter (Markowski & Richardson, 2010).

Markowski and Richardson (2010) mention that the convective clouds associated with lake-effect precipitation can be several kilometres deep. However, even shallower lake-effect clouds can produce significant amounts of snowfall (see Fig. 10), and these shallow yet intense clouds present challenges to the design of a radar network in coastal areas in a cold climate.

The organization of convection in a lake-effect snowstorm depends on the ratio of the wind speed to the maximum fetch distance. When the wind is strong, offshore convection is rapidly organized into horizontal convective rolls. When the wind is weaker, it is more likely that bands parallel to the shoreline (and perpendicular to the mean wind) are formed in the land-breeze convergence zone. With very weak winds, convection can be organized into vortices that stay over the sea and have the structure of a miniature hurricane (Laird et al., 2003). These vortices provide another example of mesoscale weather systems which can only be observed with remote sensing instruments.

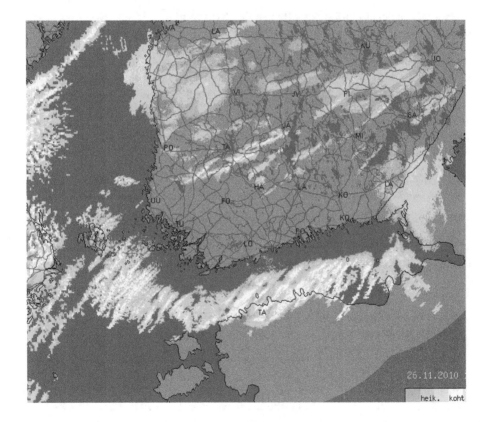

Fig. 10. Lake effect snow 26.11. 2010 10 UTC. Pseudo-CAPPI reflectivity composite of 5 Finnish and 2 Swedish radars at nominal height of 500 m.

For radar-based nowcasting applications, lake effect snow is a challenge firstly, because the snow storms do not move, thus motion vectors can be misleading, and secondly because their shallow and intense nature can cause beam overshooting problems..

7. Operational applications

Snowflakes are beautiful and interesting, but most people who buy radars want also to do something useful with the data. The most frequently asked questions are when, where and how much. These have been solved with more and less advanced accuracies over the years. Advanced questions to be still researched are related to properties of snow, most of all its density.

7.1 Accumulated snowfall, Z/S

When we talk of precipitation, we often think about rainfall, and ignore snow. There are two main approaches to accumulated snow: 1) snow water equivalent (SWE): how large is the mass of snow per unit area and 2) the (increase of) thickness of snow layer (TSL). SWE is important for hydrological applications, and it is also usually the parameters measured at surface station rain gauges (in manual gauges, the snow is melted and volume of resulting water is measured). SWE has also applications in estimates of snowload of buildings and tree crowns. Thickness of snow layer has applications in road maintenance, biology and recreational activities, and it is a tricky thing to estimate for everyone, not just radar meteorologists. A bulk equation TSL=10*SWE (both TSL and SWE expressed in mm), is often used, even though we all know that the density of snow on ground varies a lot. Matters are further complicated if we try to accumulate TSL over a longer period, because snow on ground is changing shape, density and even location (by blowing snow).

The radar equation (see Zrnic, this volume, for details) includes the parameter $|K|$, dielectricity of scattering particles, which has different values for ice and water. In operational signal processing, this is assumed to be always the water-value, so we should call the measured parameter Ze (where the "e" stands for "equivalent"). The error caused by assuming the same dielectricity is compensated in using a different ZS relation for snow, and including the effect of dielectiricity there.

The density of snow crystals and aggregates varies as a function of structure from 50 to 900 kg m^3, with higher values expected for solid ice structures and wetted particles. The size distributions of ice crystals and snow aggregates can be represented by exponential and gamma functions, and the total number of concentrations is on the order of $1-10^4$ m^{-3} for aggregates, $10-10^9$ m^3 for individual crystals at colder temperatures (T < -20 °C), and often as high as 10^4 m^{-3} at warmer temperatures (Pruppacher & Klett, 1996). The diameter of large crystals can be up to 1-5 mm, while the diameter of aggregates can grow to 20-50 mm, occasionally even more. The shapes of aggregates vary from approximately spherical to extremely oblate, and the approximate shapes of crystals can vary from extreme prolates and oblates to essentially spheres (Pruppacher & Klett, 1996). Most individual crystals tend to fall with their largest dimension horizontally oriented unless there are pronounced electric fields. Aggregates also can fall in a horizontally oriented manner or may tumble (Straka, 2005).

Radar estimates of ice water content of crystals and aggregates are greatly complicated by the multitude of crystal sizes and shapes, various crystal and aggregate densities, and dielectric constants, among others. Of all the snow types, the determination of the amount of wet aggregates is probably the most difficult (Straka, 2005).

For liquid precipitation, the classical ZR relation was published by Marshall and Palmer 1948. For snow, similar classical paper is probably that of Sekhon and Srivastava (1970). However, when dropsize distributions are known to vary a lot, snow particle distributions vary even more. Applying correction for vertical profile of reflectivity before the ZR relation is crucial, but does not eliminate all uncertainties.

7.2 Nowcasting snow

The simplest application of radar images for nowcasting is to display a time series as an animation, and visually follow its speed and direction of movement. Second level of complication is to estimate the future movement with some vector field, which can be derived from observed movement, NWP, or even Doppler velocity field (note this is not recommended but some people do it). Compared to summertime convective precipitation, snow has some advantages and some disadvantages in this respect. Because snow is seldom related to convection, it has less diurnal variation, and hence frontal snowstorms can sometimes be tracked and extrapolated for several hours with fairly good accuracy. On the other hand, snowstorms tend to be shallow, and hence the geometrical factors can cause error in speed estimates, and even causes of total miss (snowstorms hiding under the lowest radar measurement).

Because snowflakes fall slowly, they can advect remarkable distances after the radar measurement. If we measure at height of 800 m, and the snowflakes fall 1 m/s, they reach the ground 800 seconds later, and if wind blows 10 m/s, the location can be 8 km downwind from the radar measurement. From height of 1800 m, the flakes fall for half an hour, and from height of 4 km more than an hour, and for a distance in order of 40 to 60 km. This affects all studies comparing radar measurements to "ground truth", and it can be annoying for nowcasting, too. On the other hand, for an optimist it is a source of information: basically, we have already measured the snowflakes which will fall e.g. to the runway half an hour later. Lauri et al. (2012) have discussed the effect of advection in snowfall measurement.

7.3 Visibility in snow

Aviation meteorology uses abbreviation LVP (low visibility procedure) and we often read this as "fog and stratus"). However, even snowfall reduces visibility in significant amounts. Unlike in fog, the visibility in snowfall often fluctuates rapidly and significantly, and hence use of radar data to aid nowcasting would be beneficial.

Visibility is related to scattering of visible light, radar reflectivity is related to scattering of microwaves. In case of particles in typical sizes of snowflakes and snow crystals, these two behave differently: if the amount of snow in air stays same, but crystals join to larger aggregates, radar reflectivity grows (following the ND^6 equation) while the optical visibility improves (scattering gets smaller). Hence, any reflectivity-to-visibility –equations depend heavily on particle size distribution.

Rasmussen and Cole (2002) have given to visibility the equation $Vis = k / ND^2$, where k is a constant related to snow type. Having the two equations available, it would be tempting to "just solve the k" and get a reflectivity – visibility equation. However, the "constant" k can get plethora of values depending on the crystal type, the degree of riming, the degree of aggregation, and the degree of wetness of the crystals. Rasmussen et al (1999) derived ratios of visibility and liquid equivalent snowfall rate for 27 crystal types and two aggregate types. In their study, typical variations in visibility for a given liquid equivalent snowfall rate ranged from a factor of 3 to a factor of 10, depending on the storm.

After this, the next attempt would be try to "calibrate" the factor k for each storm. However, Rasmussen et al. (1999) also noted that k has a wide degree of scatter also during a given storm. As we know from other studies, the type of snow crystals depends on temperature and humidity it has experienced during its growth time. Snowstorms are often related to weather situations (such as warm fronts) with strong gradients of temperature and humidity. Hence, changes of crystal type during a storm are natural.

Another factor making comparison of reflectivity and visibility is the illumination. In same snowfall intensity, visibility in night (how far can you see a light source) can be twice as good as in daylight.

8. Conclusion

The four main properties a radar meteorologist should remember about snow are:

- It falls from shallow clouds, so you can't see it from far.
- It falls slowly, so it may advect after we measure it.
- It has different scattering properties from rain, so your precipitation estimates may be inaccurate if ZR relations are applied for snowfall
- Snowflakes come in many sizes and shapes, and their scattering properties may vary, so more information may be acquired using dual polarization.

9. Acknowledgment

The author wishes to thank Aulikki Lehkonen for finding the illustrative weather situations for examples, and Tuomo Lauri and Pekka Rossi for their constructive comments to the manuscript.

10. References

Barthazy E. & Schefold, R. (2006). Fall velocity of snowflakes of different riming degree and crystal types. *Atm. Res.*, 391-398

Boodoo, S., Hudak, D. Donaldson, N. & Leduc, M. (2010). Application of dual polarization radar melting-layer detection algorithm. *J. Appl. Meteor. Clim.,* 49, 1779–1793.

Bringi V. N. & Chandrasekar, V. (2001). *Polarimetric Doppler weather radar: principles and applications,* Cambridge University Press. ISBN 978-0521623841

Giangrande, S., Ryzhkov, A.V. & Krause J. (2005). Automatic detection of the melting layer with a polarimetric prototype of the WSR-88D radar. 32nd Conference on Radar Meteorology, Amer. Meteor. Soc., Albuquerque, NM, available online via
http://ams.confex.com/ams/32Rad11Meso/techprogram/paper 95894.htm.

Koistinen, J., Michelson, D.B., Hohti, H. & Peura, M. (2003). Operational measurement of precipitation in cold climates. *Weather Radar: Principles and Advanced Applications,* pp. 78–110. P. Meischner, Ed., Springer, ISBN 978-3540003281 Berlin Heidelberg.

Laird, N. F., Walsh, J. E. & D. R. Kristovich, D.R. (2003). Model simulations examining the relationship of lake-effect morphology to lake shape, wind direction, and wind speed. *Mon. Wea. Rev.,* 131, 2102–2111.

Lauri, T., Koistinen, J. & Moisseev, D. (2012). Advection based Adjustment of Radar Measurements *Mon. Wea. Rev.,* 140, 1014-1022.

Markowski, P. & Richardson, Y. (2010). *Mesoscale Meteorology in Midlatitudes.* John Wiley and Sons, ISBN 978-1119966678, Chichester, UK.

Nakaya U. (1954). *Snow Crystals: Natural and Artificial.* Harvard University Press. ISBN 978-0674811515. Cambridge, UK.

Nicosia, D. J. & Grumm R. H. (1999). Mesoscale Band Formation in Three Major Northeastern United States Snowstorms. *Wea. Forecasting,* 14, 346–368.

Pruppacher, H. R. & Klett J.D. (1996). *Microphysics of Clouds and Precipitation.* Second ed., Springer, ISBN 978-0792342113. Heidelberg London New York.

Rasmussen, R. & Cole, J.A (2002). How Snow Can Fool Pilots. National Center for Atmospheric Research, Boulder, CO, USA. Available online at
http://www.rap.ucar.edu/projects/wsddm/SNOFOOL.pdf

Rasmussen, R. M., Vivekanandan, J., Cole, J. Myers, B. & Masters, C (1999): The Estimation of Snowfall Rate Using Visibility. *J. Appl. Meteor.,* 38, 1542–1563.

Ryzhkov, A.V. & Zrnic, D.S. (1998). Discrimination between rain and snow with a polarimetric radar, *J. Appl. Meteor.,* 37, 1228-1240.

Ryzhkov, A. V., Schuur T.J., Burgess, D.W., Heinselman, P.L., Giangrande, S.E. & Zrnic, D.S. (2005). The Joint Polarization Experiment: Polarimetric Rainfall Measurements and Hydrometeor Classification. *Bull. Amer. Meteor. Soc.,* 86, 809–824.

Sekhon, R. S., & Srivastava, R. C. (1970). Snow size spectra and radar reflectivity. *J. Atmos. Sci.,* 27, 299–307.

Straka, J., Zrnic, D.S. & Ryzhkov, A. (2000). Bulk hydrometeor classification and quantification using polarimetric radar data: Synthesis of relations. *J. Appl. Meteor.,* 29, 1341–1372.

Straka J., & Zrnic, D.S. (1993). An algorithm to deduce hydrometeor types and contents from multiparameter radar data. *Preprints, 26th Conf. on Radar Meteorology, Norman, OK, Amer. Meteor. Soc.*, 513–516.

6

Use of Radar Precipitation Estimates in Urban Areas: A Case Study of Mexico City

Ernesto Caetano[1], Baldemar Méndez-Antonio[2] and Víctor Magaña[1]
[1]Instituto de Geografía, National Autonomous University of Mexico,
[2]Energy Department, Metropolitan Autonomous University,
Mexico

1. Introduction

Storm events have long been a menace to Mexico City. The main reason is related to the fact that in summer, many showers can reach intensities of more than 20 mm/hour, which makes difficult the management of the drainage system in various areas of the city. As instance discharges of large magnitudes in the western part of the city are an element of danger, as they lead to flash flood that inundate populated areas downstream in a matter of minutes. Recent flooding events in Mexico City have revealed its vulnerability to severe weather conditions. Although regularization programs and new urban land policies are been implemented by the council government, there are still many families living in high risk areas. These areas over hillsides, and irregular human settlements still proliferate. Usually, severe storms can cause hazard landslides because unstable landfills and deforested hill slopes. On the other hand in the flat parts of the city, faulty drainage systems usually cause sewage flooding after continuous rain events. The urban sprawl undergone in the last half century, has not kept pace with urban services such as drainage. In the rainy season, puddles arise, sometimes caused by the presence of silt and debris in the ducts and the drainage system capacity is exceeded, and in other cases there are no absorption wells in areas with problems in the drainage network. Additionally, the lack of maintenance of dams and channels can also result in severe flooding problems. In most cases, the intense rainfall events produce merely an emergency response of fire departments.

In the last three decades there have been major advances in remote sensing techniques for estimation of rain, mainly in the use of meteorological radar and weather satellites, increasing the availability of rainfall data for operational meteorological and hydrological applications. Precipitation estimates derived from meteorological radar are useful in runoff simulation in urban drainage. Spatial distribution of radar rainfall used as input to a distributed hydrological model permit to characterize the performance of drainage infrastructure at local and regional scale. Radar data used in this analysis are obtained from C-band radar deployed at western Mexico Valley basin and derived rainfall estimates provides the input to a distributed hydrologic model applied to the Mixcoac microwatershed located at western Mexico basin. Radar and distributed hydrologic model are capable to provide accurate rainfall and runoff data supporting specific-site flood information and, also provides a baseline for comparison and guides design of radar network as one component of an early warning system for the region.

The international practice aims at a comprehensive approach to flood management in response not only to the consequences of a specific event, but measures that starting from the prediction of extreme events and monitoring for early warning purposes to establishment of civil protection measures for those affected by the occurrence of such events and the hazard remediation, including infrastructure development and non-structural measures to reduce the vulnerability.

The structure of this work is as follows. Section 2 presents the summer precipitation regime over Mexico Valley. Section 3 a brief description the meteorological radar deployed in Cerro Catedral at western of Mexico Valley. In Section 4 a case study for the microwatershed of the Mixcoac River is discussed. In Section 5 new design of a weather information system is proposed. Concluding remarks are found in Section 6.

2. Mexico valley precipitation climatology

The México Valley is located at 2240 m altitude and at a latitude of approximately 19°N and is characterized by well-defined rainy season from late May to early October which can be classified as a monsoon climate type.

The orography of region plays an important role on the precipitation patterns (intensity, timing, spatial distribution and, extreme events occurrence). On northeastern area, region nearly flat, the average precipitation is around 500 mm/year and at southwestern mountainous region part of the México Valley, the average reaches almost 1200 mm/year (Fig. 1). The occurrence of extreme events follows the same patterns as shown by Magaña et al. (2003) by establishing a criterion to determine when intense precipitation should be considered an extreme event based on a Gamma distribution of the observed amount of daily rainfall, for each station of the rain gauge network (Fig. 2). The similarity of spatial variability of severe weather and average precipitation becomes apparent. An extreme precipitation event to occur in the western or southern part of the basin, rainfall in 24 hours should exceed 25 or 30 mm, while in the eastern part of the city, more than 15 mm in 24 hours already constitute an extreme event. To a large extent the interaction of the mountains with the summer easterly winds determines, the characteristics of precipitation (Barros 1994).

The summer precipitation diurnal cycle indicates the intense precipitation begins in the afternoon (Fig. 3), around 16:00 h local time, generally in the eastern part of the valley, and propagate to the western part during the evening, reaching the other extreme of the valley by late evening and midnight (Mendez et al., 2006).

3. The Cerro Catedral radar

The radar network operated by Mexico's National Water Commission is focalized mainly in monitoring of the tropical cyclone activity and consists of thirteen C-band radars (manufactured by Ericsson Inc., Enterprise Electronics Corporation and Vaisala). There is also a weather radar deployed at Cerro Catedral which cover almost all Mexico Basin (located 40 km and altitude 3785 m, approximately 1500 meter above México City), to monitor severe weather over this region (Fig. 4). This radar measures reflectivity and has a Doppler and dual polarization and, is configured as follows (Table 1):

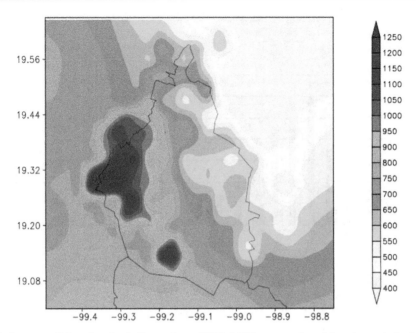

Fig. 1. Summer (May-October) climatology (2003-2008) accumulated (mm) precipitation distribution over Mexico Valley.

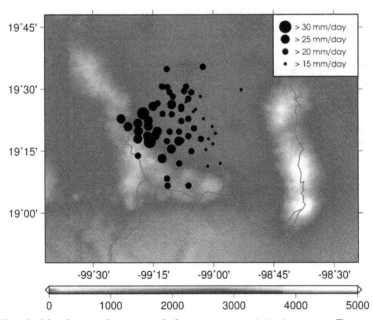

Fig. 2. Threshold values to determine daily extreme precipitation event. Topography (m) is shown in color.

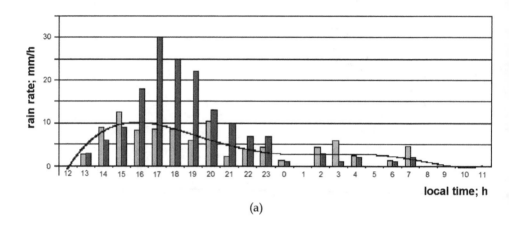

(a)

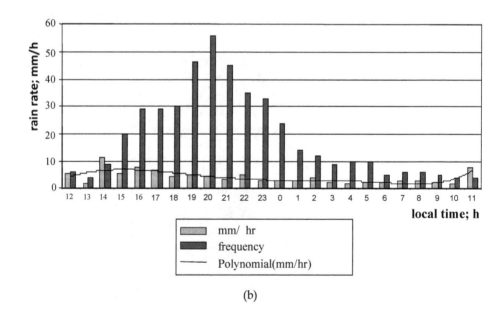

(b)

Fig. 3. Precipitation intensity, frequency and occurrence time histograms; (a) northern and (b) southern Mexico basin; (1993-2001).

Antenna diameter (m)	4.2.
Antenna Gain (dB)	44.7
Beamwith (")	0.9
Polarization	Linear hor/vert
Frequency (GHz)	5.60-5.65
Wavelenght (cm)	5.30-535
Peak power (kW)	250
Pulse lenght (µs)	0.5 - 2.0
PRF (Hz)	250, 900, 1200
MDS (dBm)	-114, -110, -109

Table 1. Cerro Catedral weather radar (Ericsson UBS 103 04, upgraded by Sigmet/Vaisala technology) technical characteristics.

Such elevation has effects over the precipitation estimated at low level on Mexico Valley Watershed. In order to get a good coverage of shallow rain, originating close to ground, it is necessary to settle the elevation angles to negative value of around –1.5 degree. This has the inconvenience of blockage, clutter and loss of signal. The weather radar usually suffers partial or total blockade operating in mountain zones due to the complex topography around it. This effect can limit the coverage of the radar when it use negative degrees and affect the precipitation measurements (Joss & Waldvogel, 1990; Sauvageot, 1994; Collier, 1996 and Smith, 1998). The application of some blockage corrections to the observations radar is worthwhile, in order to get quantitative estimation of the precipitation and it can be combined with elimination of spurious echoes by two and three-dimensional analysis of the topography and the storm (Krajewski &Vignal, 2001; Steiner & Smith, 2002). A promising development in this field is related to the gradual change of weather radar concept, from a tool for qualitative rainfall estimation to a tool for more quantitative rainfall measurement (Borga et al., 1997).

Despite the drawback of the radar height site (1500 m above Mexico City) to follow stratiform precipitation system, convective systems are adequately monitored. In fact, the radar is capable of doing the full scan within convective clouds but no precipitation estimates in clouds with a base height of less than 3500 m (Fig. 5).

Figure 6 shows the monitoring of three storm events within range of radar coverage. In addition to tracking the storms, which would support decision makers in a warning system in these figures one can see the fixed echoes caused by the presence of volcanoes on the eastern side of Mexico City. This represents a serious problem in estimating precipitation, both qualitatively and in its distribution and location, as it provides information on areas where there is rain. Considering that the fixed echoes, whether caused by the interception of the land or effect of the lateral lobes, they can be largely eliminated with the Doppler radar function. However, despite all the Mexican radar, including that of Cerro Catedral, have this feature is not used to remove these echoes.

The Doppler radar function is a great help to eliminate this kind of echo and it is important for a better estimation of rainfall fields for hydrological or/and alert purposes. The cost of sending false warning alerts to users, when these echoes are not removed, is high, because once lost confidence in warnings of severe storms is difficult to recover it.

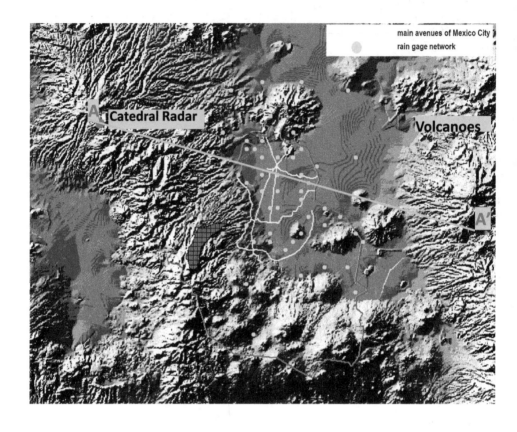

Fig. 4. Mixcoac River Basin (pink shaded area), Cathedral radar site and main avenues of Mexico City and rain gauges network.

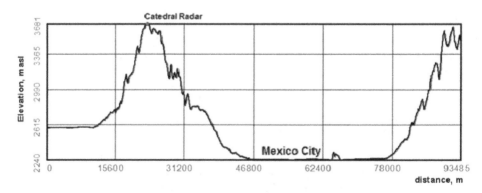

Fig. 5. Relief cross section A-A' of the Mexico Valley basin (see Fig. 4).

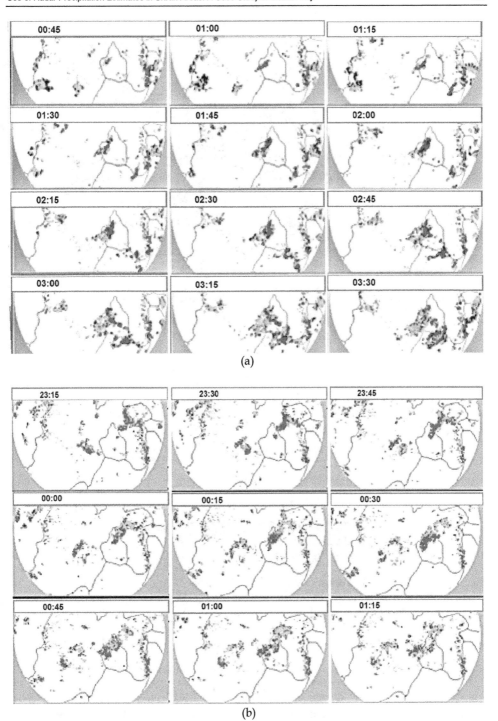

(a)

(b)

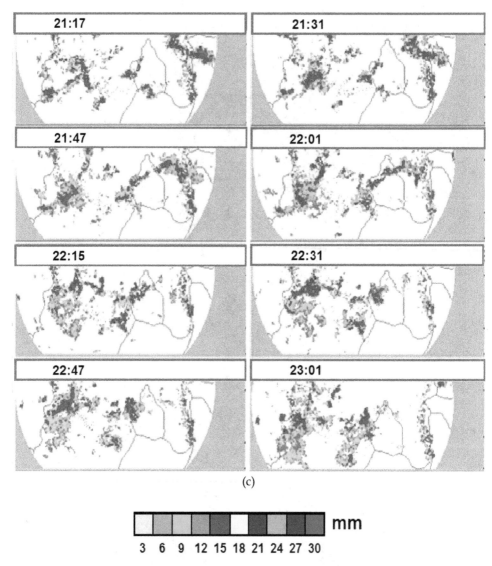

(c)

Fig. 6. Examples of storm radar monitoring over the Valley of Mexico basin and identification of fixed echoes caused by the interception of the radar beam with volcanoes located eastern Mexico City: a) 15 June; b) 13 July and; c) 19 September 1998.

4. Case study

Mendez (2005) and Mendez et al. (2009) using reflectivity radar data for the period of 1995-1998, selected 13 intense precipitation events to examine the rainfall patterns over México City, determined by the Cerro Catedral radar, by looking at spatial characteristics (shape, position and magnitude) of precipitation across the valley. The analysis also constitutes a first step towards an improved understanding of storms over urban areas, particularly during the summer rainy season.

Méndez et al. (2011) developed a lumped model of the rainfall-runoff type with input from radar and pluviograph (rain gauge network of Mexico City water management system) data for calibration, applied to the microwatershed of Mixcoac River located at western Valley of Mexico basin, over an area of 31.5 km² (Fig. 4).

Currently all Mexican radar precipitation estimatives use the Marshall-Palmer equation (Marshall and Palmer, 1948), which presents certain discrepancy for tropical regions (Fig. 7). However, the precipitation underestimation could also be associated to beam overshooting, attenuation or hardware calibration issues. A local calibration was performed in order to improve the rainfall estimation taking into account characteristics of precipitation system over Mexico (Mendez et al., 2006). The resulting local calibration improves the estimation of rain (Fig. 7). Although false echoes treatment is not yet done, one might think that it can improve even more if they are removed.

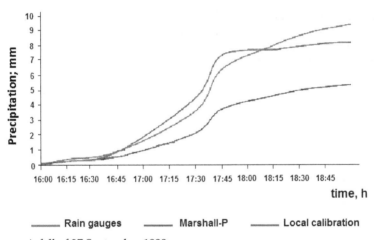

Fig. 7. Storm rainfall of 27 September 1998.

It should be mentioned the radar estimates reproduce well the precipitation patterns in time but not quantitative ones. This new calibration distribution improves this estimate.

4.1 Hydrological analysis

The improvement in the unitary hydrograph of the basin is clear (Fig 8). This is an expected result if one notices that the radar initially reproduce the temporal variability of rainfall and subsequently, after calibration hydrology, quantitative estimate improves. It is obvious that the radar properties would be underused in aggregated hydrological models because the ability to detect the precipitation spatial variability usually are not take into account in these models and, thus the analysis of the hydrological processes within the basin. However, an analysis of the aggregated model applied here is performed to detect the advantage of the radar data for hydrological model in an experimental urban catchment.

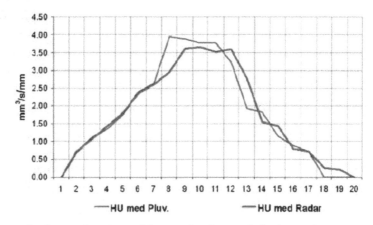

Fig. 8. The unit hydrograph estimated from radar data and pluviographs.

Additionally, in order to obtain a methodology to determine the distributed parameter hydrologic model in other watersheds, the experimental basin of the river Mixcoac and the technique Distributed Unit Hydrograph (Clark, 1943) is used. The conceptual model obtained (Fig. 9) is similar to Maidment model (Maidment, 1993). The model obtains the isochrones to Mixcoac River Basin (Fig. 10), which is then used to estimate the outflow hydrograph of the basin. The comparison between the observed and estimated with the distributed hydrological model fed with precipitation data obtained from the radar is showed in the figure 11.

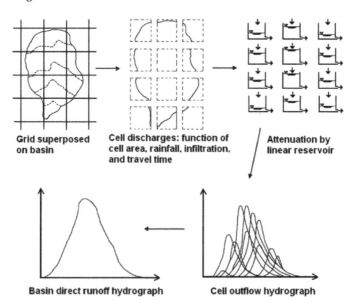

Fig. 9. Conceptual model of the distributed hydrologic model known as the Modified Clark (Source: Kull & Feldman, 1998).

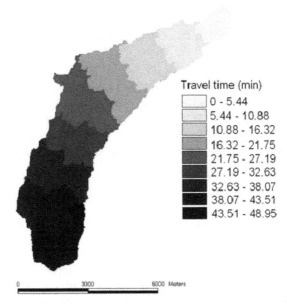

Fig. 10. Mixcoac watershed Isochronous.

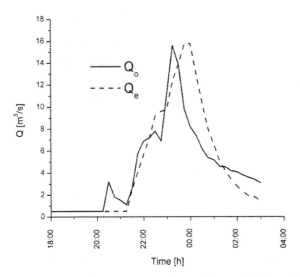

Fig. 11. Hydrological response of Mixcoac river basin, observed (Q_o) and estimated (Q_e) with radar rainfall data.

The observed and estimated response using radar rainfall data demonstrate a fairly agreement and creates the confidence to apply the methodology used in this analysis, in other watersheds. It is important to establish how to estimate correctly the radar rainfall data by eliminating false echoes, using Doppler mode techniques to filter those echoes caused by the interception beam (principal and/or secondary lobes) with the terrain. This will give greater confidence to hydrologists that the data used were carefully treated before feeding to their models.

Aiming to use the radar hydrological information with operational purposes, a short hydrological forecast system should include the following components integrated components: a) Automatic meteorological stations network (pluviographs); b) Weather Radar; c) Satellite products and; d) Mesoscale numerical weather prediction model. The main feature of each of these components is to provide real time and hours/days in advance (forecast) rainfall data, which is a necessary condition for implementing an operational hydrological system.

5. Weather forecast system

In most cases, the fire department response facing intense rainfall events over Mexico basin is a merely emergency procedure. Its work would be greatly improved and lead to more efficient use of human and material resources available by the city government by taking advantage of weather information that is, diagnoses and forecasts of weather and climate. Unfortunately, the information prepared by the National Weather Service lacks the detail and quality required for making decisions as presented in general terms, without data, in order to one can acquire confidence in the forecast. This problem is particularly severe when it comes to prediction of severe storms considered as a danger to the water system in Mexico City. Requirements to take the first steps in the right direction are the improvement of surface measurement networks, radar and satellite information, forecasting deadlines to produce hydrometeorological information useful in decision making for disaster prevention and development of an early warning system that includes not just the danger or threat, but also the vulnerability facing to severe weather.

Any centre that generates meteorological information for decision making is based on the following required elements (Fig. 12):

1. Data Collection
2. Assimilation and display of information. Very short-term prognosis based on radar and satellite estimates of rainfall and rain gauge information.
3. Weather Forecast Systems
4. Post-processing of weather forecasting in the short term to prepare products tailored to user needs
5. Scheme for submission of information to the user or decision maker, including an early warning system useful for the Water System of Mexico City
6. Seasonal climate forecasts for water management in the long run

Although all components are equally important, the main focus to be discussed here is the radar network to be implemented in the system. Méndez et al. (2009) presented a proposal for a new radar system for Mexico Valley based on precipitation analysis estimated by the existing Cerro Catedral radar. They found an underestimation in the amount of precipitation over the western mountains of the valley, at the foothills, while rainfall rates tend to be overestimated over the eastern parts, resulting of the blockage effect of mountains between the current radar position and the basin. The weather radar usually suffers partial or total blockade when it is located in mountainous regions. The current location of the Cerro Catedral radar can limit the coverage of the radar when it requires negative angles in the vertical measurements to monitor the valley, affecting the precipitation estimates (Joss and Waldvogel, 1990; Sauvageot, 1994; Collier, 1996; Smith, 1998).

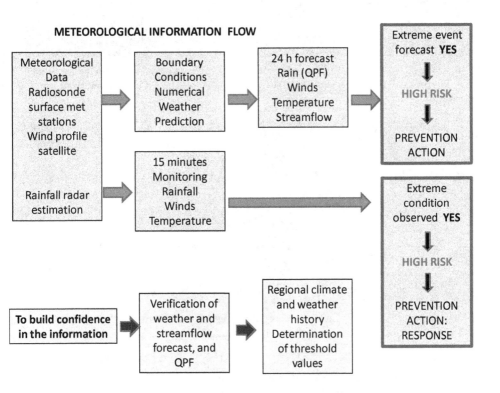

Fig. 12. Meteorological information flow of Early Warning System.

In order to achieve a complete three-dimensional coverage of México City a second radar deployment in the opposite extreme of the basin is required. Several conditions of propagation either from the present radar or from others possible positions (Fig. 13a) were attempted to get the greater coverage area (Méndez et al., 2009). The site selected was the Cerro de la Estrella located at the central eastern of Mexico Valley (Fig. 13b) had shown more adequate. The Cerro de la Estrella is at an approximate elevation of 300 m above the City of Mexico and therefore is able to scan both stratiform and convective precipitation (Fig. 14). The radar coverage was obtained from a Geographic Information System.

Fig. 13. a) Proposed sites in Mexico Basin for the new radar deployment. Black line is the political boundary of Mexico City: b) View of the Valley of Mexico basin from the Cerro de la Estrella.

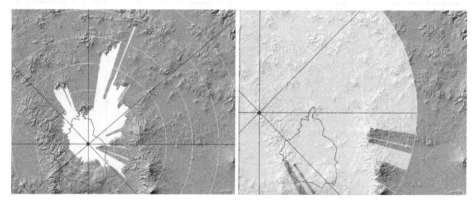

Fig. 14. Radar scan at 0 deg.: Left, The Cerro de La Estrella (2450 msl) and; right the Cerro Catedral site (3785 msl). Red line is the political boundary of Mexico City.

The Mexico Basin rain gauge network (Méndez et al. 2009) is very dense (Fig. 4) and the new Mexico Valley radar system should take advantage of this to implement quantitative precipitation estimation schemes as one of main products generated by the early warning system. To achieve this, a proper processing of radar data must implemented in order to develop methodology to prevent beam blockage due to orography (Bech et al. 2003), ground clutter (Fornasiero et al., 2006), which can produce frequent false alarms and affect the precipitation estimative. These effects can be mitigated through the application of the decision-tree method proposed by Lee et al. (1995) for a dual-polarized system, which able to provide additional parameters such as differential reflectivity, correlation coefficient (and their texture) that can be used to further reinforce the traditional techniques.

All products generated by the early warning system (graphic, data image, text, bulletin) will be integrated into a display system based in GIS system. This will permit produce better quality graphic resolution and generate tailor made products for specific needs to stake holders and general public.

6. Conclusion

Hydrological applications of meteorological radars have become an important branch of remote sensing in meteorology and disaster preparedness activities. The high temporal and spatial resolution precipitation fields generated by meteorological radars meet the requirements of the hydrological modeling (Sempere-Torres et al., 2004). Furthermore the radar covers large areas and is of rapid access for real time hydrological applications and, therefore an adequate blending of radar and rain gauge data results in better estimates of real time precipitation (Collier,1996; Joss & Waldvogel, 1990). Méndez et al. (2009) has pointed the capacity of the Doppler radars to scan storm gives a big advantage for runoff and precipitation prediction and may be fundamental to understand the physics of storm intensification in complex orography as México Valley.

The system of weather/hydrological forecasting and monitoring storms enable user and stake holders to have information of more severe events in advance and establish risk management policies for their mitigation. The scheme for dissemination of information must contemplate to present the results of diagnostic scheme and weather forecasting as clearly as possible in order that users and stake holders have relevant elements to incorporate objective vulnerability assessments to more closely meet the facing risks. To achieve this the continuous results display in a color system associated with critical values of risk, using a Geographic Information System, is a powerful tool for prevention and response prevention or emergency in accordance with the Mexico City government interests. Further improvements in the short term precipitation forecast, or quantitative precipitation forecast (QPF), can be achieved by blending Doppler radar products and output of numerical weather prediction models (Atencia et al. 2010)

The results of this study and Méndez et al. (2011) assess the value of using weather radar data as input distributed hydrological models. These models are adequate for applications in regions of strong slopes and heavy rainfall with complex draining networks for which reason it would be very useful in early warning systems.

Early warning systems, widely used in the world, aim to provide relevant information for making decisions within a framework of prevention. This type of action has proved to be much more helpful, even under weather forecasts uncertainties, than a system based only an emergency response.

7. References

Atencia, A.; Rigo, T., Sairouni, A., Moré, J., Bech, J., Vilaclara, E., Cunillera. J., Llasat. M. C., & Garrote, L. (2010). Improving QPF by blending techniques at the Meteorological Service of Catalonia. *Nat. Hazards Earth Syst. Sci.*, 10, 1443–1455.

Barros, A. P. (1994). Dynamic modeling of orographically induced precipitation. *Rev. Geophys.*, 32, 265-284.

Bech, J.; Codina, B., Lorente, J., & Bebbington, D. (2003). The Sensitivity of Single PolarizationWeather Radar Beam Blockage Correction to Variability in the Vertical Refractivity Gradient. *J. Atmos. Ocean. Technol.*, 20, 845–855.

Borga, M.; Da Ros, D., Fattorelli, S., & Vizzaccaro, A. (1997). Influence of various weather radar correction procedures on mean areal rainfall estimation and rainfall-runoff simulation. *Weather radar technologies for water resources management*. Braga, B. J. and

Massambani, O. (Eds.), IRTCUD/University of Sao Paulo, Brazil and IHP-UNESCO, 73-86.

Clark, C. O. (1943). Storage and the Unit Hydrograph. *Transactions of the American Society of Civil Engineers*, 110, 1419-1446.

Collier, C. G. (1996). *Applications of weather radar systems*. John Wiley and Sons, 2nd Ed., New York, 390 pp.

Fornasiero, A.; Alberoni, P. P., & J. Bech (2006). Statistical analysis and modelling of weather radar beam propagation conditions in the Po Valley (Italy). *Nat. Hazards Earth Syst. Sci.*, 6, 303-314.

Joss, J., & Waldvogel, A. (1990). Precipitation Measurement and Hydrology. *Radar in Meteorology*, D. Atlas, Ed.. Amer. Meteor. Soc., 577-597.

Krajewski, W. F., & Vignal, B. (2001). Evaluation of anomalous propagation echo detection in WSR-88D data: a large sample case study. *J. Atmos. Ocean. Tech.*, 18, 807-814.

Kull D. W. & Feldman A. D., (1998) Evolution of Clark's Unit Graphs Method to Spatially Distributed Runoff. *Journal Hydrology Engineering*, ASCE 3 (1), 9-19.

Lee, R.; Della Bruna, G., & Joss, J., (1995). Intensity of ground clutter and echoes of anomalous propagation and its elimination. *Proc. On the 27th Conference on Radar Meteorology*, the Amer. Meteor. Soc., Vail, Colorado, 651-652.

Magaña, V.; Pérez, J., & Méndez, M. (2003). Diagnosis and prognosis of extreme precipitation events in the México City watershed. *Geofísica Int.*, 41, 247-259.

Maidment, D.R. (1993). GIS and Hydrologic Modeling. *Environmental Modeling with GIS*, ed. by M.F. Goodchild, B.O. Parks, and L.T. Steyaert, Oxford University Press, New York, pp. 147-167.

Marshall, J.S., & Palmer, W.M. (1948). The distribution of raindrops with size. *Journal of Meteorology*, 5 , 165-166.

Méndez, B. (2005). *Aplicación hidrológica de los radares meteorológicos*. PhD. Thesis, Faculty of Engineering - UNAM, 186 pp. (*in Spanish*).

Méndez, B.; Domínguez, R., Magaña, V., Caetano, E., & Carrizosa, E. (2006). Calibración hidrológica de radares meteorológicos. *Ing. Hidrául. Méx.*, 21, 43-64. (*in Spanish*).

Méndez, B.; Magaña, V., Caetano, E., Silveira, R., & Domínguez, R. (2009): Analysis of daily precipitation based on weather radar information in México City. *Atmosfera*, 22(3), 299-313.

Méndez, B.; Domínguez, R., Rivera-Trejo, F., Soto-Cortés, G., Magaña, V., & E. Caetano, (2011). Radars, an alternative in hydrologic modeling. Aggregate Model. *Atmosfera*, 24(2), 157-171.

Sauvageot, H. (1994): Rainfall measurement by radar: A review. *Atmos. Res.*, 35, 27-54.

Sempere-Torres, D.; Corral, C., Sánchez-Diezma, R., Berenguer, M., Velasco, C., Franco, M., Llort, X., Velasco, E., & Pastor, J. (2004). Are radar rainfall estimates ready for hydrological applications? Some reflections from the experience in Catalunya. *European Conference on Radar in Meteorology and Hydrology (ERAD) – COST 717*. Final Seminar. Abstracts Book, 30 p.

Smith, P. L., Jr. (1998): On the minimum useful elevation angle for weather surveillance radar scans. *J. Atmos. Oceanic Technol.*, 15, 841-843.

Steiner, M.; & Smith, J. A. (2002). Use of three-dimensional reflectivity structure for automated detection and removal of nonprecipitating echoes in radar data. *J. Atmos. Ocean. Tech.*, 19, 673-686.

7

A Network of Portable, Low-Cost, X-Band Radars

Marco Gabella[1,2], Riccardo Notarpietro[2], Silvano Bertoldo[3], Andrea Prato[2], Claudio Lucianaz[3], Oscar Rorato[3], Marco Allegretti[3] and Giovanni Perona[3]

[1]*Meteoswiss*
[2]*Politecnico di Torino – Electronics Department,*
[3]*Consorzio Interuniversitario per la Fisica delle Atmosfere (CINFAI) – Sede di Torino*
[1]*Switzerland*
[2,3]*Italy*

1. Introduction

1.1 Excellent qualitative overview of the weather in space and time

Radar is a unique tool to get an overview on the weather situation, given its high spatio-temporal resolution. Over 60 years, researchers have been investigating ways for obtaining the best use of radar. As a result we often find assurances on how much radar is a useful tool, and it is! After this initial statement, however, regularly comes a long list on how to increase the accuracy of radar or in what direction to move for improving it. Perhaps we should rather ask: is the resulting data good enough for our application? The answers are often more complicated than desired. At first, some people expect miracles. Then, when their wishes are disappointed, they discard radar as a tool: both attitudes are wrong; radar is a unique tool to obtain an excellent overview on what is happening: when and where it is happening. At short ranges, we may even get good quantitative data. But at longer ranges it may be impossible to obtain the desired precision, e.g. the precision needed to alert people living in small catchments in mountainous terrain. We would have to set the critical limit for an alert so low that this limit would lead to an unacceptable rate of false alarms.

1.2 Range dependence of the results (range degradation)

Perhaps accurate quantitative precipitation estimation (QPE) can only be achieved at short ranges from the radar. This is not because we miss careful investigations, but simply, because radar can only see the hydrometeors aloft, while we would need to know what is arriving at ground level. Obstacles as well as earth curvature lead to a limited horizon, allowing us to see precipitation at variable height, often too far from the ground. All these difficulties increase rapidly with range from the radar location. The situation becomes obviously much more difficult in mountainous terrain, where weather echoes can only be detected at high altitudes because of beam shielding by relieves: there, terrain blockage combined with the shallow depth of precipitation during cold seasons and low melting levels causes inadequate radar coverage to support QPE, especially in narrow valleys.

Furthermore, precipitation is too variable for the "coarse" resolution of long-range ground-based radars (GR). The variability of natural precipitation is so large that the radar beam often does not resolve it. As a result we find aloft different types of particles and non-homogeneous reflectivity in the pulse volume, to be compared with rain rate at the ground level. The under-sampling problem becomes increasingly severe with increasing ranges because the radar backscattering volume increases with the square of the range; therefore, at longer ranges, small but intense features of the precipitation system are blurred (non-homogeneous beam filling). Furthermore, it is more likely to include different types of hydrometeors (e.g., snow, ice, and rain drops), especially in the vertical dimension. We know that, on average, the radar backscattered echo from liquid, mixed phase, and frozen hydrometeors decreases with height. Using several TRMM overpasses, the comparison between the TRMM radar and linearly averaged GR radar reflectivity, carried out in circular rings around the GR site, has clearly confirmed a significant range dependence of the TRMM/GR ratio (*Gabella et al.* [2006], *Gabella et al.* [2011a], *Gabella et al.*. [2011b]). This well-known problem is caused mainly by the increasing sampling volume of the long-range GR with range, combined with non-homogeneous beam filling: e.g., at longer ranges of GR, the lower part of the volume could be in rain, whereas the upper part of the same pulse can be filled with snow, ice, and mixed phase particles. Quite often it can be even characterized by an echo weaker than the radar sensitivity itself (apparently, no backscattered echo). This phenomenon (called "beam overshooting" by radar meteorologists) is also caused by the decrease of vertical resolution with range, thus amplifying the influence of the horizon and Earth's curvature. Because of beam overshooting, strong range degradation has been noticed in several parts of the world when analyzing weather radar data over a long time period. The reader can refer, for instance, to the 2-year analysis by *Young et al.* [1999] in the United States or by *Gabella et al.* [2005] in the Swiss Alps.

In mountainous terrain, precipitation is even more variable both in space and time because of orographic effects and interactions of mountains with wind fields. This variability within the scattering volume is in contradiction with the homogeneously filled pulse volume assumption usually made when considering the meteorological radar equation. Fulfilling the assumption of homogeneous beam filling, however, is a prerequisite for a precise estimate of reflectivity, attenuation and phase shift along the beam.

1.3 Type and width of the distribution of precipitation

Another fundamental problem is the asymmetry and the large variability of precipitation rates in time and space. In other words, distributions are wide and skewed-to-the-high-end at the same time. This statement concerns particle type, particle size, number density of particles as well as derived integral parameters such as reflectivity, rain- and snow-rate. As a consequence of the distributions in time and space, we find that a small area (say 1/10 of the "rainy" area, which in turn can be 1/20 of the surveillance area ...) during a "short" time (i.e. smaller than the rainy/cloudy period) contributes a large fraction of the total precipitation amount. As a direct consequence of this (small "time/space" of significant and heavy rain rate), the chance of detecting weak rain rate is much larger than high rain rate. Without careful thinking and without having analyzed large data sets, we may be tempted to extrapolate the rules of weak rain into strong one. This extrapolation will involve large errors, because mechanisms producing rain vary with its intensity. In other words, different mechanisms produce weak and large rain rates.

1.4 Difficulties with conventional long-range radar: Inability to observe the lower part of the troposphere combined with non-homogeneous beam filling

We may wonder: why is it so difficult to grasp a realistic precision out of "long-range" (say two hundred kilometers) weather radar? Perhaps, the main reason can be found in the difficulty of reproducing the results verified with large effort at close ranges. We cannot extrapolate them to the full range displayed by our operational, meteorological radars. At short ranges problems caused by shielding, inhomogeneous beam filling, attenuation and vertical profile may be dealt with. This is not possible at longer ranges. This statement does not exclude the use for weather forecasting in full range of our radars. The radar tells us where and when something is coming; radar data are helpful to validate the forecasts of the Numerical Weather Prediction models. Here, combining the information of many radars into a network may help a lot. The combination of data from many radars may also mitigate the effects caused by the range-dependence of each single radar.

Long-range radar networks remain an essential part of the weather forecasting and warning infrastructures used by many nations worldwide. Despite significant capability and continuous improvement, one fundamental limitation of today's weather radar networks is the inability to observe the lower part of the atmosphere and detect fine-scale weather features. Designed for long-range coverage through precipitation, these radars must operate at radar wavelengths not subject to attenuation. This implies the use of large antennas (to achieve narrow beam width) and high-power transmitters (to meet sensitivity requirements at long ranges); up to now, such large antennas are mechanically scanned, hence requiring dedicated land, towers and other support infrastructures. Consequently, the installation and acquisition cost of each site is usually much larger than the cost of the sensor itself.

1.5 Proposed solution: Distributed networks of many, inexpensive, redundant, low-cost, high temporal resolution, short-range, small radars

How to tackle the emerging need for improved low-altitude coverage, high temporal-resolution meteorological radars? Many low-cost, fast-scanning, short-range X-band radars for rain monitoring can be a valid solution for complementing long-range radars. Long range radars have proved to be useful for weather forecasting and qualitative surveillance. As already discussed in Sec. 1.2, the results, verified with large effort at close ranges, cannot be generalized. Because of range degradation (non uniform beam filling and overshooting, see Sec. 1.2), it seems impossible to reproduce the results easily obtained close to the radar for quantitative applications at far ranges. This is especially true in mountainous terrain. Therefore, an interesting solution could be to combine the data of many, small, low-cost and short-range X-band radar for rain estimates within valleys.

2. The potential for distributed networks of small low-cost weather radars

2.1 The work of the remote sensing group at the Politecnico di Torino

The European INTERREG IIIB Alpine Space Programme started in 2004 the FORALPS Project ("Meteo-hydrological Forecast and Observations for improved water Resource management in the ALPS"). One of the aims of FORALPS was the design and development of a portable, low-cost, small radar for weather monitoring. The Remote Sensing Group at the Politecnico di Torino was involved in the development activities of this new network starting from its early ideation stages (*Notarpietro et al.* [2005]).

The first designed scenario was specifically intended to cover narrow valleys within the Alps. This was initially achieved by adopting a non-conventional vertical plane scanning strategy with a fan beam slot waveguide antenna (1° beam width in the vertical plane, 25° beam width along the valley). The initial implementation was simply designed to collect two low elevation acquisitions with opposite directions along the valley plus a vertical sounding to evaluate the vertical reflectivity profile (*Gabella et al.* [2008]). Then, this initial approach was extended to collect radar sounding coming from the entire vertical plane. This kind of small low-cost radar has been patented and is now sold with the name "wind-mill" mini-radar.

In a second stage, the more conventional horizontal scanning strategy was implemented to cover wide planar areas with very high temporal resolution at a fixed, optimized elevation. This suggested combining a number of short-range, low-cost radars into a network concept, to obtain a set of similar small unattended units, tightly connected within a unified environment. The result of the above approaches and suggestions is an unmaned, low-consumption, network of low-cost, small, X-band radars. Adding up, the first prototypes, running since October 2006, were installed on the Politecnico di Torino roof, sensing either the horizontal or the vertical planes. During these years several progresses and modifications were made, leading to a network of mini radars: one operated by the Aosta Valley Civil Protection (since March 2007) and a vertical scanner unit (wind-mill) installed next to the glide path of the "Sandro Pertini" Turin International Airport. Recently (autumn 2010), four horizontal scanners units were installed in different areas of Sicily (see the web site http://meteoradar.polito.it/). At present, seven small radars have been installed on the Italian territory and are successfully running. In our approach, such network is capable of mapping storms with temporal resolution better than 1 min and focusing on the low-troposphere "gap" region. Such network has the potential to complement the long-range radar networks in use today. In Chapters 3 and 4 the deployment of small, low-cost, X-band radars will be presented for the following environments:

- heavily populated areas (e.g. Palermo town and harbor, see Sec. 3.2 and 4.2; Turin town, see Sec. 4.4);
- specific dry and semi-arid regions where it is crucial to improve observation of low-level meteorological phenomena (e.g. western Sicily, Sec 4.3)
- deep valleys surrounded by high mountais region (e.g. Valle d'Aosta, see Sec. 3.1).

2.2 X-band, "short" wavelength technology for short-range monitoring

Cost, radiation safety issues and aesthetic issues motivate the use of small antennas and low-power transmitters that could be installed on either low-cost towers or existing infrastructures such as rooftops of existing buildings or telecommunication poles. This requires that the radars are physically small and that the radiated power levels are low enough so as not to pose an actual or perceived radiation safety hazard. We have opted for a very small parabolic antenna (D = 0.6 m) which corresponds to a 3 dB cross-range spatial resolution of 1 km at 20 km range (two third of the used range, which is 30 km). The antenna is hidden below a 1 m diameter radome (Fig. 1, left picture) and rotates at ~120° per second using a single elevation.

One precipitation map is made available every minute by averaging 16 rotations (out of the 22 available) 9 consecutive rays and 2 range-bins, hence resulting in a total of 144 samples.

Table 1 summarizes main characteristics of our low-cost weather radars for both configurations described in Sec. 2.1 and 2.2: the innovative "wind-mill" (tailored to narrow valleys in mountainous terrain) and the more conventional horizontal scanner (also called "super-gauge"). More details on the temporal sampling scheme and averaging process are given in the next Section 2.3.

Frequency	X-Band (9.4 GHz)
Range	up to 30 km
Power	10 kW
Pulse duration for long (short) pulse	400 (80) ns
Pulse Repetition Frequency	800 (3200) Hz; for long (short) pulse, respectively
1] Super-Gauge	Single elevation, 3.6° beam width, horizontal plane
2] Wind-mill	1.2° along valley, 25° across valley, vertical plane
Antenna (depending on scanning)	1] 0.6 m paraboloid; 2] slotted waveguide
Cost of a mini radar	< 30 kEuro (within a network of, say, 6 radars)

Table 1. Main characteristics of the low-cost weather radar.

2.3 High temporal resolution

The use of precipitation estimates from weather radar has been limited not only by the quantitative accuracy but also by the spatio-temporal resolution: firstly, there is a significant number of sources of uncertainty in the process of converting the reflectivity volume data measured by a radar to an estimate of falling precipitation close to the ground. The factors that contribute to this uncertainty have been introduced and summarized in Section 1. Secondly, the spatio-temporal resolution of radar-based QPE products from weather radar networks was generally insufficient, especially for small-scale hydrological applications. Hence, one important advantage of our mini-radar with small antenna approach is also the high temporal resolution.

Fig. 1. The portable, low-cost weather radar with (left picture) and without (right picture) the 1-meter Diameter radome. The configuration shown here is the so-called "super-gauge", which is a single-elevation scanning in the horizontal plane with a 3.6° beam width parabolic reflector (Diameter of the paraboloid is 0.6 m).

As described in Sec. 2.2, both kinds of mini radar, the "wind-mill" (vertical plane scanning along valley) and the "super-gauge" (which implements the traditional horizontal scan) currently deliver an image of precipitation every minute; furthermore, the temporal resolution can easily be reduced to 30 or even 15 s.

We here describe in more details how the time-averaged (1-minute-sampled) radar reflectivity measurements are acquired. For each of the 9 consecutive shots, 2 contiguous pulses have been acquired: the 2 contiguous pulses are separated by the pulse width, which is 400 ns; the 9 consecutive shots are separated by the pulse repetition interval, which is 1.25 ms. Every minute, the antenna performs 22 revolutions; however only data from the first 16 revolutions (out of 22) were averaged on a linear power scale (algebraic average: dBm values are antilog transformed, then averaged, then again transformed on a decibel logarithmic scale). This means a total of 288 (18 times 16) samples; among them, at least 2×16 are independent, if we assume a decorrelation time of ~10 ms (Fig. 1.14, *Sauvageot* [1992]): sample #1 and # 9 of the 9 consecutive rays are in fact separated by 9×1.25 = 11.25 ms.

3. A few qualitative examples

3.1 An hostile environment: Detecting precipitation even inside a narrow valley

At the beginning of November 2011 (from 3 to 9 November around noon) six days of continuous, wide-spread precipitation hit the north-western part of Italy (see Fig. 2). In the south-western Alps and in the surrounding flatlands and hills, in fact, autumn is the season in which the longest and heaviest rainfalls occur. This fact has long been known: a description of these "late-summer" Mediterranean storms can already be found in the works by the old-Roman author, Plinius. It can been explained in simple terms as follows: in autumn the Mediterranean Sea surface temperature is still high, while cold air is already forming over the central-northern part of Europe. This has two effects: first of all, the thermal contrast facilitates the deepening of pressure low over the north-western part of the Mediterranean Sea; secondly, the warm air that arrives from the south, flowing over the Mediterranean, provides a ready source of moisture.

The enforced rising of this warm-humid convectively unstable air, thanks to the Alpine barrier, causes extensive and heavy rainfall. One has the impression of being subject to a long storm, but, in reality, it is the continuous formation of stormy cells over the same place.

The first study site here presented is located in north-western Italy in the "Aosta Valley", which is the smallest region in Italy. It is set between the Graian and Pennine Alps, which are very steep. Among the more than "four-thousand" massifs, the most famous are: Mont Blanc, Monte Rosa, the Matterhorn and Gran Paradiso. The Dora Baltea river together with its tributaries have formed the tree-leaf-shape veining of the Valley. A Digital Elevation map of the investigated area is shown in Fig. 2.

Being surrounded by such high relieves (> 4000 m MSL), the deep Aosta Valley (< 500 m MSL) cannot be effectively monitored by any of the surrounding weather radars (Dole, close to Geneva; Bric, close to Torino; Monte Lema, close to Maggiore Lake). Among these three radars, the one with "less worse" visibility is certainly Monte Lema, which was the only one able to detect some weak echoes during the 24-hour period shown in Fig. 3a (from 12 UTC of November 4 to 12 UTC of November 5). However, because of beam shielding by relieves combined with overshooting, the 24-hour radar-derived rainfall amounts above the central-western part of the Aosta Valley are heavily underestimated: for instance, above Aosta town, the Swiss weather radar network (see Fig. 3a) shows amounts smaller than 2 mm in 24 hours.

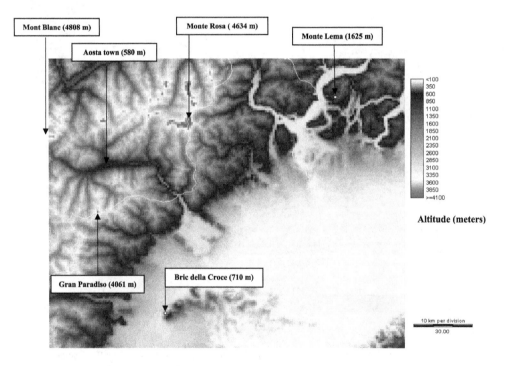

Fig. 2. Digital Elevation map of the north-western part of Italy.

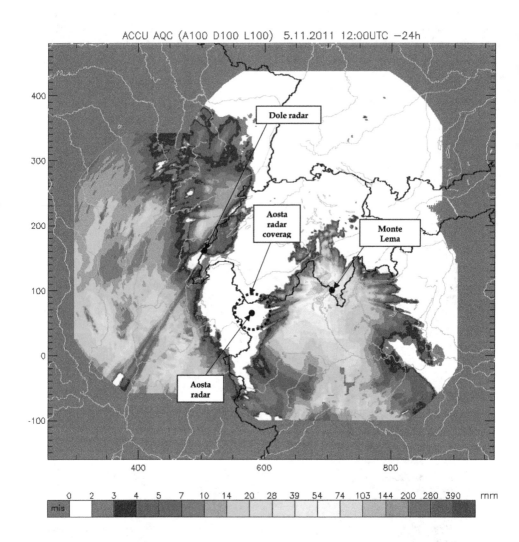

Fig. 3a. 24-hour cumulative rainfall amounts in the western Alps as seen by the Swiss weather radar network (from 12 UTC of November 4 to 12 UTC of November 5, 2011). North and East axes map units are in km.

What if we supplement long-range weather radar information with precipitation fields derived at high spatio-temporal resolution by portable, low-cost X-band radars? The answer is given in Fig. 3b, which shows what the low-cost X-band radar can detect, despite being deployed down deeply into the valley. As it can be seen, the 24-hour cumulative precipitation amounts surrounding Aosta town indicate values between 16 and 25 mm in 24 hours (yellow patch). According to rain gauges, such amount still represent ~2 dB radar underestimation: from 12 UTC of November 4 to 12 UTC of November 5, in fact, the gauges "Aosta Piazza Plouves" (580 m MSL) and "Aosta St. Christophe" (550 m MSL) respectively measured 44.2 mm and 40.2 mm.

Finally, it is worth noting that the very complex orography causes severe beam shielding: the radar is practically blind at all ranges in the northern part of the circular surveillance area while in the southern half-circle weather echoes are only detected in approximately 10% of the 30 km range.

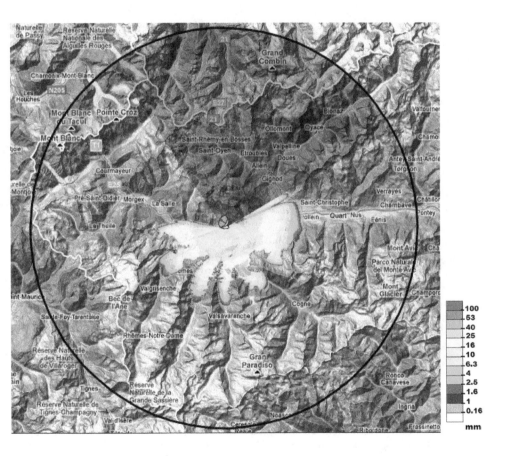

Fig. 3b. 24-hour cumulative rainfall amounts in the north-western part of Italy as seen by the low-cost X-band radar located near the town of Aosta (from 12 UTC of November 4 to 12 UTC of November 5, 2011). The circular range ring is at 30 km range from the site of the mini-radar.

3.2 The extreme spatio-temporal variability of the precipitation field in semi-arid regions

In this section, a typical Mediterranean thunderstorm hitting the Palermo town in Sicily is presented. More details are given in the caption of Fig. 4 below, while QPE performances of the Palermo mini radar are thoroughly discussed in Sec. 4.2.1 and 4.2.2.

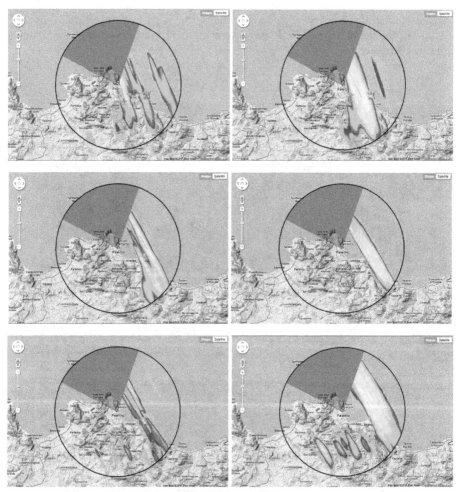

Fig. 4. Average hourly precipitation field on the evening of February 18, 2011 in the northern part of a dry Mediterranean island (Sicily); it is worth noting the "wide precipitation band" shape and the high spatial variability of the field despite the averaging process used to derive hourly cumulated rainfall amounts. Each of the 6 consecutive pictures shows the average of 60 instantaneous maps of radar reflectivity (one per minute) transformed into equivalent rain rate using a fixed Z-R relationship. The first picture shows hourly accumulation rainfall amounts from 16 to 17 UTC; the last one from 21 to 22 UTC. The circular range ring is at 30 km range from the site of the mini-radar.

4. Some quantitative examples in Sicily and Piedmont

4.1 Quantitative precipitation estimation (QPE)

While Section 3 dealt with qualitative examples, in the present Section 4 we will present hourly radar-derived precipitation amounts as obtained from weather echoes aloft to be compared with point rainfall measurements acquired at the ground by rain gauges.

4.1.1 From instantaneous radar reflectivity to hourly rain rate amounts

We have seen in Sec. 2.3 that the mini-radar finally provides an instantaneous radar reflectivity value once per minute for each radar bin of 3° by 120 m. This value is in turn the average of 288 samples (among them, at least 32 samples are independent, see Sec. 2.3).

It is well known that the backscattered power caused by rain drops is, unfortunately, only indirectly linked to the rain rate, R ([R] = mm/h). The backscattered power caused by the hydrometeors and detected by the radar is, in fact, directly proportional to the radar reflectivity factor, Z. A fundamental quantity for precise assessment of both Z and R is the drop size distribution (DSD), N(D), which is defined as the number of rain drops per unit volume in the diameter interval δD, i.e. between the diameter D and $D+\delta D$. The radar reflectivity factor, Z, is defined as the 6th moment of the DSD, namely:

$$Z = \int_0^\infty N(D)D^6 dD \cdot \tag{1}$$

In radar meteorology, it is common to use the dimensions of mm for drop diameter, D, and to consider the summation (integral) to take place over a unit volume of 1 m³. Therefore, the conventional unit of Z is in mm⁶/m³. For the assessment of rain rate, another fundamental quantity is needed: the terminal drops fall velocity as a function of the diameter, v(D). Since it is common to use [v] = m/s, then the relationship is

$$R = 6 \cdot 10^{-4} \cdot \pi \int_0^\infty N(D)D^3 v(D)dD \cdot \tag{2}$$

If precipitating hydrometeors in the radar backscattering volume were all spherical raindrops (which is almost never the case!) and the DSD could be described to a good approximation by an exponential DSD, then a simple power-law would relate Z to R. The first ever exponential DSD presented in a peer-reviewed paper and probably the most quoted is the Marshall-Palmer (M-P) distribution. The power law derived using the exponential fit proposed in Eq. (1) and (3) of the famous paper by *Marshall and Palmer* [1948] is Z=296·R^1.47.

Here we have used the following Z-R relationship Z = 316·R^1.5 to derive the variable of interest, R, from the geophysical observable, Z, which is detected by the meteorological radar. Such values have been retrieved by *Doelling et al.* [1998] using seven years of measurements in central Europe. It is also worth noting that for the 2 radars in Sicily prior to any processing, the radar reflectivity values were increased by 4 dB to compensate system losses not properly compensated in the "traditional" radar equation.

For each radar bin, a maximum of 60 clutter-free radar reflectivity values are then transformed into R using $R = (Z/316)^{2/3}$ and then averaged to derive the corresponding hourly rain rate used in this study.

4.1.2 QPE evaluation based on the comparison between hourly radar and gauge rainfall amounts

The evaluation is based by looking at the average value and the dispersion of the errors (we call error the disagreement between radar and gauge amounts). For such characterization, we define the two following parameters:

1. *Bias* (in dB). The bias in dB is defined as the ratio between radar and gauge total precipitation amounts on a logarithmic (decibel) scale. It describes the overall agreement between radar estimates and ground point measurements. It is averaged over the whole space–time window of the sample. A positive (negative) bias in dB denotes an overall radar overestimation (underestimation).

2. *Scatter* (in dB). The definition of scatter is strictly connected to the selected error distribution from a hydrological (end-user) and radar-meteorological (operational remotely sensed samples of the spatio-temporal variability of the precipitation field) perspective. The error distribution is expressed as the cumulative contribution to total rainfall (hydrologist point of view, y axis) as a function of the radar–gauge ratio (radar-meteorologist point of view, x axis). Most of the sources of error in radar precipitation estimates, in fact, have a multiplicative (rather than an additive) nature. An example of the error distribution is shown in Fig. 2 and 3 of *Germann et al.* [2004]. The scatter is defined as half the distance between the 16% and 84% percentiles of the error distribution. The scatter refers to the spread of radar–gauge ratios when pooling together all volumetric radar estimates aloft and point measurements at the ground.

From our radar-meteorological point of view the multiplicative nature of the error prevails with respect to the additive one. For example, water on the radome, a wrong calibration radar constant, or a bad estimate of the profile all result in a multiplicative error (i.e. a factor) rather than an additive error (i.e. a difference). This is why bias, error distribution and scatter are expressed as ratios in dB. A 3 dB scatter, for instance, means that radar–gauge ratios vary by a factor of 2. If bias is zero, it is interpreted as follows: the radar-derived estimate lies within a factor of 2 of the gauge estimate for 68% of rainfall while for the remaining 32% the uncertainty is larger. The scatter as defined above is a robust measure of the spread. It is insensitive to outliers for two reasons. First, each radar–gauge pair is weighted by its contribution to total rainfall (y axis of the cumulative error distribution). An ill-defined large ratio that results from two small values, e.g. 0.4 mm/2 mm ~ −7 dB, describes an irrelevant event from a hydrological point of view, and only gets little weight. Second, by taking the distance between the 16% and the 84% percentiles, the tails of the error distribution are not overrated. Another important advantage of the spread measure is that it is unaffected by the bias error, hence providing a complementary view of the error in the estimates. The above definition of the scatter is thus a better measure of the spread than the less resilient standard deviation.

4.2 Quantitative precipitation estimation for the Palermo radar

The Palermo radar is located on a small hill next to the harbor of the capital of Sicily (blue triangle, Fig. 5): its latitude is 38°.1139; its longitude is 13°.358; its altitude is 45 m above Mean Sea Level (MSL). The radar has been installed in autumn 2010.

For the quantitative evaluation of the radar estimates, the most reasonable available rain gauge in terms of range and radar visibility is the one located in Altofonte. It is run by the Servizio Informativo Agrometeorologico Siciliano (SIAS). Its location (red triangle) is shown in Fig. 5, which represents a 90 m resolution Digital Elevation Model of the region at two different scales: left picture domain is ~ 540 by 540 km²; right picture is 90 by 90 km².

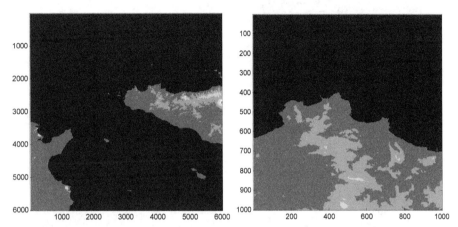

Fig. 5. Digital Elevation Map of Sicily showing two different domains (pixel size is 90 m); the blue triangle shows the site of the mini radar next to Palermo down town. The red triangle shows the location of the most reasonable rain gauge in terms of range and radar visibility (see Fig. 6); this gauge (Altofonte) has been used for the QPE evaluation. Axes map units are 90-m pixels.

The radar-gauge distance is 11.1 km; the rain gauge altitude is 370 m above MSL. Fig. 6 shows the radar-gauge profile as derived using the DEM shown in Fig. 5. In addition to the terrain profile (black curve), the picture shows the mini-radar 3.6° Half Power Beam Width (HPBW, often called "3 dB beamwidth" in radar meteorology) by means of two blue lines. The radar beam axis, which divides such angular sector in two equal parts of 1.8°, has an angle of elevation equal to +3°. As it can be seen from Fig. 6, the gauge location is not optimal: in the last kilometer before the gauge, the 3 dB portion of the primary lobe hits the hilly terrain, hence causing some beam shielding (power loss) and ground clutter contamination. However, regarding this last problem, it is worth noting that the rain gauge location is behind the top of the hill: this means that Palermo radar echoes above the Altofonte gauge are practically ground-clutter-free; nevertheless, as stated, because of beam shielding, some (radar) underestimation above the gauge can be expected.

For what concerns QPE, 6 rainy days (144 hours) during the first 4 months of 2011 have been analyzed; these days are February 1, 23 and 28, March 5, April 26 and 27. During these 144 hours the Gauge (Radar) total amounts was 77.4 (62.7) mm, which corresponds to an "overall Bias" of –0.9 dB (radar underestimation). Out of 144 analyzed hours, in 48 (42) cases the Gauge (Radar) derived hourly rainfall amount was larger than 0.4 mm/h, which is the hydrological threshold adopted in this Chapter for discriminating between "hydrologically speaking" wet and dry hours (see also Table 2).

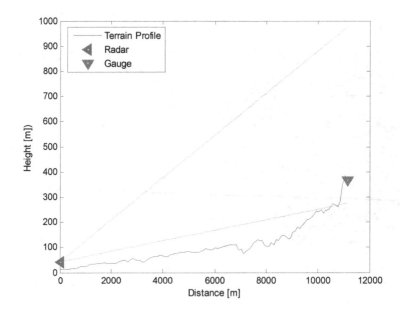

Fig. 6. Vertical section of the terrain profile from the Palermo radar site to the Altofonte gauge derived from the Digital Elevation Model shown in Fig. 5. The blue lines indicate the boundaries of the radar antenna Half Power Beam Width.

The scope of our QPE analysis is twofold:

- to evaluate from an hydrological point of view the radar ability in discriminating wet versus dry hours;
- to assess quantitatively the radar accuracy in estimating hourly rain rates.

As introduced in the methodological Section 4.1.2, the latter quantitative assessment will be based on the Scatter in dB and thoroughly described in Sec. 4.2.2; the former evaluation will be presented in the Sec. 4.2.1.

4.2.1 Wet versus dry hours discrimination according to radar echoes using the gauge as reference

The history of applying contingency tables (also called error matrices in the remote sensing field) for the verification of one set of observations against a reference set is a quite long one. The history of categorical statistics based on such tables is rather fascinating and an interesting account is given by Murphy (1996). Most of the scores were first derived nearly a century ago and have been rediscovered several times (with different names in different branches of science, see for instance the bullet list below).

The dimension of the contingency tables can be as small as 2×2 (tetrachoric) or larger (polychoric) depending on the number of thresholds used in the classification scheme. Obviously, in our wet-versus-dry hourly values discrimination, we are dealing with tetrachoric tables, since just one discrete value (namely 0.4 mm/h) is used to divide the two categories. On the one hand, the properties of a set of observations can be condensed and clearly displayed through such tables; on the other hand, to satisfy needs of specific users, even for a simple tetrachoric table, several different scores have been introduced. Here, we will use two scores that can be applied to both tetrachoric and polychoric tables:

- the Heidke Skill Score (HSS), also known as Kappa Index of Agreement (KIA), Khat, …
- the Hanssen-Kuipers (HK) score also known as True Skill Score (TSS), …

Details regarding these two scores can be found in literature; the interested reader may refer, among others, to the paper by *Tartaglione* [2010] regarding HK and to the work by *Hogan et al.* [2009] regarding HSS. In particular, the Appendix of this last paper interestingly aims at estimating confidence intervals in the HSS.

	G ≥ 0.4 mm/h	G < 0.4 mm/h	
R ≥ 0.4 mm/h	38	4	42
R < 0.4 mm/h	10	92	102
	48	96	144

Table 2. Contingency table between the Palermo radar estimates and the Altofonte rain gauge measurements.

From Table 2, it is straightforward to derive the Probability Of Detection (POD), which is 38/48= 0.79 and the False Alarm Ratio (FAR), which is 4/42= 0.095.

Regarding the two above mentioned multi-categorical scores selected, **HSS** results to be **0.774**; HK is slightly smaller: 0.750.

4.2.2 Quantitative agreement between gauge and radar-derived hourly rain rates

During the 48 "Wet-Gauge" hours, the total rainfall amount according to the Altofonte Gauge was 72.4 mm. According the Palermo Radar, the total rainfall amount was 55.9 mm, which means a residual "Wet-Gauge Mean Field Bias" of −1.1 dB (radar underestimation). Based on these 48 **"Wet-Gauge"** hourly amounts, which are shown in Fig. 7, the **Scatter** results to be **2.38 dB**, as can also be seen from the 48 values displayed in Fig. 8.

Using only the 38 hours where BOTH the Radar AND the Gauge amounts were larger or equal to 0.4 mm/h, the total Gauge (Radar) amount was 64.6 (54.7) mm. Consequently, the "wet-wet" Mean Field Bias is −0.7 dB. As expected, also the "wet-wet" Scatter improves: based on such 38 "wet-wet" hourly amounts, it results to be 1.97 dB, as can be seen from Fig. 9. The 0.4 dB decrease in the value of the Scatter is a clue of the not-negligible rain amount missed by the radar during the 10 "Missing Detection hours".

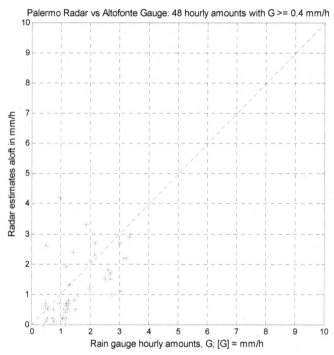

Fig. 7. Scatter plot of 48 hours with Altofonte gauge amounts larger or equal than 0.4 mm/h and the corresponding Palermo radar estimates. The scale is linear and the maximum value is set to 10 mm/h (as in Fig. 16 for the Torino site).

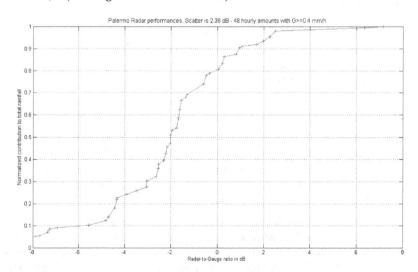

Fig. 8. Cumulative contribution to total rainfall as a function of Radar-Gauge ratio for hourly precipitation with $G \geq 0.4$ mm/h (48 samples).

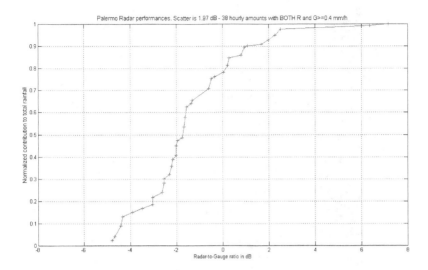

Fig. 9. Cumulative contribution to total rainfall as a function of Radar-Gauge ratio for hourly precipitation with both R and G $\geq$ 0.4 mm/h (38 samples).

4.3 Quantitative precipitation estimation for the Bisacquino radar

The Bisacquino radar is located in the central-western part of Sicily: latitude is 37°.707; longitude is 13°.262; altitude is 780 m above Mean Sea Level (MSL). As for the Palermo radar (Sec. 4.2), it has been installed in autumn 2010.

For the quantitative evaluation of the radar estimates there is an optimal rain gauge run by the Servizio Informativo Agrometeorologico Siciliano (SIAS) and installed in the municipality of Giuliana (località Castellana). Its location (red triangle) with respect to the radar (blue triangle) is shown in Fig. 10 together with the DEM of the area.

The radar-gauge distance is 8.7 km; the rain gauge altitude is 250 m above MSL. As can be seen from Fig. 11, the gauge location is optimal not only in terms of range, but most of all in terms of radar visibility: no partial beam shielding by relieves affects the mini radar 3.6° Half Power Beam Width (HPBW), which is delimited by the two blue lines in Fig. 11 (the radar beam axis has an angle of elevation set to 1°).

Furthermore, the hill located at 3 km range from the radar, causes partial beam shielding of the remaining part of the primary lobe and total shielding of the secondary lobes in elevation; consequently, we can conclude that residual ground clutter contamination affecting radar echoes above the Giuliana rain gauge is negligible.

For what concerns QPE, 4 rainy days (96 hours) during the first 5 months of 2011 have been analyzed; these days are February 23, March 13, April 26 and May 22. During these 96 hours the Gauge (Radar) total amounts was 98.2 (77.7) mm, which corresponds to an "overall Bias" of –1.0 dB (radar underestimation).

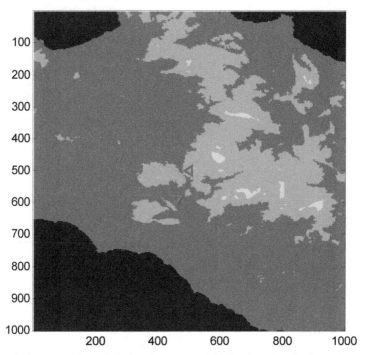

Fig. 10. Digital Elevation Map of the western part of Sicily showing the Bisacquino radar site (red triangle) and the Giuliana rain gauge location (blue triangle). Axes map units are 90-m pixels.

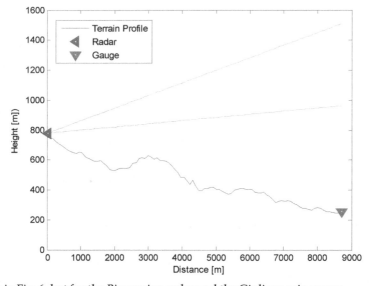

Fig. 11. As in Fig. 6, but for the Bisacquino radar and the Giuliana rain gauge.

Out of 96 analyzed hours, in 37 (34) cases the Gauge (Radar) derived hourly rainfall amount was larger than 0.4 mm/h (see the next Sec. 4.3.1 and Table 3 for more details).

4.3.1 Wet versus dry hours discrimination

From Table 3, it is easy to see that in this case the POD is (29/37) 0.78 while the FAR is relatively high: (5/34) 0.15. If we consider scores that deal with all the elements of the table, then **HSS** results to be **0.710** while HK is quite similar: 0.699.

	$G \geq 0.4$ mm/h	$G < 0.4$ mm/h	
$R \geq 0.4$ mm/h	29	5	34
$R < 0.4$ mm/h	8	54	62
	37	59	96

Table 3. Contingency table between the Bisacquino radar observations and the Giuliana rain gauge measurements during the 96-hour observation period.

4.3.2 Quantitative agreement between gauge and radar-derived hourly rain rates

During the 37 "Wet-Gauge hours", the total rainfall amount according to the Giuliana Gauge was 97.4 mm. According to the Palermo Radar, the total rainfall amount was 74.6 mm, which means a residual "Wet-Gauge Mean Field Bias" of −1.2 dB (radar underestimation). Based on these 37 "**Wet-Gauge**" hourly amounts, the **Scatter** results to be **1.47 dB**. (as can be seen in Fig. 12). Such 37 Radar-Gauge data pairs are shown in Fig. 13.

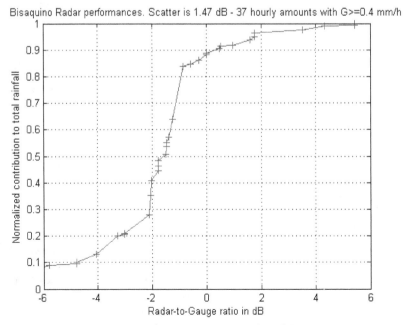

Fig. 12. As for Fig. 8, but for the 37 samples ($G \geq 0.4$ mm/h) of the Giuliana rain gauge.

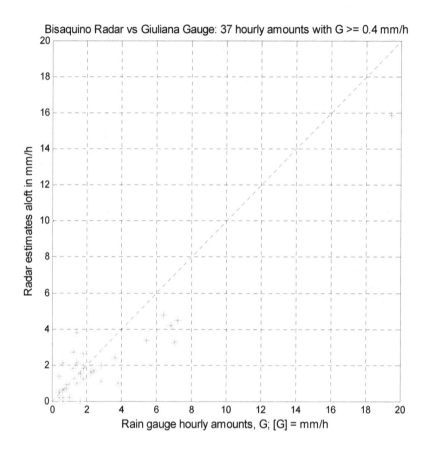

Fig. 13. Scatter plot of 37 hours with Giuliana gauge amounts larger or equal than 0.4 mm/h and the corresponding Bisacquino radar estimates. It is worth noting that in this case the maximum value of the linear scale for the scatter plot is 20 mm/h.

Using only the 29 hours where BOTH the Radar AND the Gauge amounts were larger or equal to 0.4 mm/h, the total Gauge (Radar) amount was 90.6 (73.5) mm. Consequently, the "wet-wet" Mean Field Bias is –0.9 dB. Also the "wet-wet" Scatter improves slightly: based on such 29 "wet-wet" hours, it results to be 1.31 dB, as it can be seen from Fig. 14. The 0.16 dB decrease in the value of the Scatter is a clue of the almost marginal rain amount missed by the radar during the 8 "Missing Detection hours" (6.8/97.4).

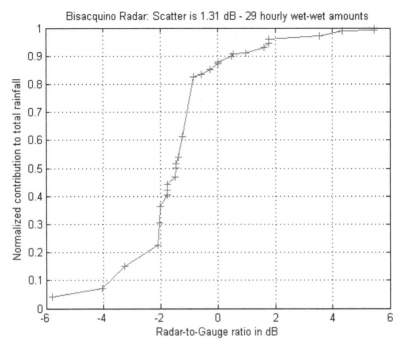

Fig. 14. As for Fig. 12, but for 29 samples with both R and G ≥ 0.4 mm/h.

4.4 Quantitative precipitation estimation for the Torino radar

The Torino radar is located on the roof of the Politecnico di Torino (Electronics and Telecommunications Department): its latitude is 45°.063; the longitude is 7°.660; the altitude is 275 m above Mean Sea Level (MSL).

For the quantitative evaluation of the radar estimates we have been using three rain gauges: Castagneto Po, Ciriè and Nichelino. The observation period is based on 42 hours.

For the rain gauge of Castagneto Po, the radar visibility is similar to that of Palermo radar versus Altofonte gauge, although the range is considerably longer: 21.1 km. As it can be seen from Fig. 15 (left picture), which shows the radar-gauge profile, the ground clutter contamination at the gauge location should be negligible since the device is luckily just behind the top of the hill: this means that Torino radar echoes above the gauge are at least ground-clutter-free. For the rain gauge in Ciriè, the range is 18.7 km, while the situation is worse in terms of ground clutter: the gauge location (Fig. 15, right) is, in fact, visible from the radar site. For the Nichelino gauges the situation is similar.

For what concerns QPE, we have at our disposal data from the 3 gauges during the same 42 hours (April 2011). During these 126 hours the Gauge (Radar) total amounts was 121.7 (114.2) mm, which corresponds to an "overall Bias" of –0.3 dB (radar underestimation).

Out of 126 analyzed hours, in 63 (45) cases the Gauge (Radar) derived hourly rainfall amount was larger than 0.4 mm/h (see the next Sec. 4.4.1 and Table 4 for more details).

4.4.1 Wet versus dry hours discrimination

From Table 4, we can see that on the one hand the number of Missing Detections (22) is relatively high: hence POD is relatively small 0.65; FAR is relatively small: 0.089. Finally, **HSS** results to be **0.587**, which is by chance exactly the same as HK .

	G ≥ 0.4 mm/h	G < 0.4 mm/h	
R ≥ 0.4 mm/h	41	4	45
R < 0.4 mm/h	22	59	81
	63	63	126

Table 4. Contingency table for the Torino radar and 3 gauges (42-hour observation period).

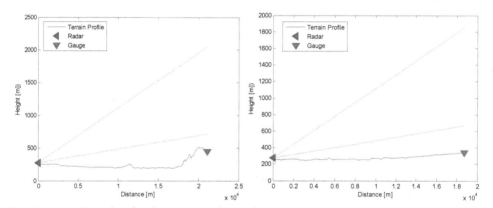

Fig. 15. As in Fig. 6, but for the Torino radar. Left: Castagneto Po-Torino Radar profile; right: Ciriè-Torino Radar profile.

4.4.2 Quantitative agreement between gauge and radar-derived hourly rain rates

During the 63 "Wet-Gauge hours", the total rainfall amount measured by the three gauges was 119.7 mm. The corresponding radar-derived amount was 108.0 mm. Based on these "**Wet-Gauge**" 63-hourly amounts the Bias is –0.4 dB. As can be clearly seen from Fig. 16, the agreement between the radar and the 3 gauges is quite poor. This fact is obviously reflected in the amazingly large value of the Scatter, which is as bad as 5.38! dB.

Fig. 17 shows the 63 "Wet-Gauge hours" hourly amounts as measured by the gauges at the ground and as derived from radar echoes aloft: the large scatter between such different devices and their different sampling modes is again evident.

Using only the 41 hours where BOTH the Radar AND the Gauge amounts were larger or equal to 0.4 mm/h, the total rainfall amount measured by the three gauges is reduced to 94.5 mm. The corresponding radar-derived amount remains instead almost the same: 106.2 mm. Consequently, the "wet-wet" Mean Field Bias increases and becomes even positive: +0.5 dB (radar overestimation). Also the "wet-wet" Scatter improves remarkably: based on such 41 "wet-wet" hours, it results to be 3.73 dB, which is still a figure much worse than the one obtained for the 2 radars in Sicily. Such huge Scatter value decrease (high sensitivity to different radar thresholds) is again a clue of the poor QPE agreement for the Torino radar.

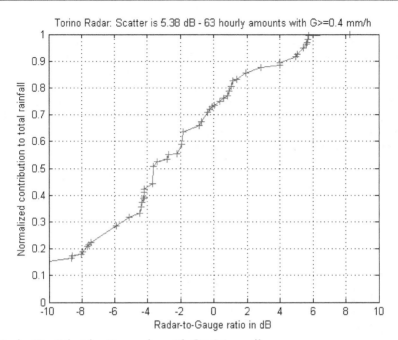

Fig. 16. As for Fig. 8, but for 63 samples with G ≥ 0.4 mm/h.

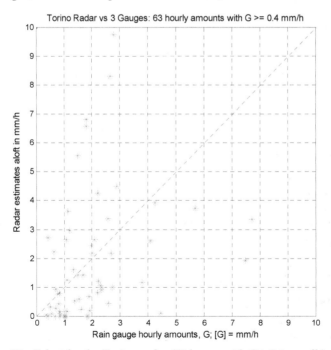

Fig. 17. Same as Fig. 7, but for the Torino radar (63 hours with G ≥ 0.4 mm/h).

4.5 QPE Summary in terms of bias and scatter in dB

For the three mini-radar sites discussed in the previous paragraphs, Table 5 provides a summary of the QPE evaluation in terms of Bias in dB (as a function of the number of considered hours). When all the wet and dry hours are considered, we have the so-called "overall" Bias: it includes False Alarm events (mainly caused by ground clutter contamination) and Missing Detection (caused by beam overshooting, see Sec. 1.2 and, in some cases, attenuation) events; it is certainly the most resistant and complete definition of Bias. It is obviously a measure of the mean error and says nothing about the error dispersion around the mean. For this purpose, there is the Scatter, which will be presented in the Table 6.

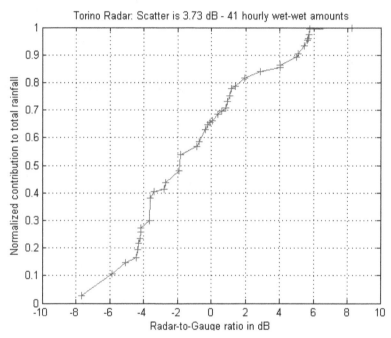

Fig. 18. As for Fig. 16, but for but for 41 samples with both R and G ≥ 0.4 mm/h.

While the "overall" Bias (1st column) includes both wet and dry periods, the other two Bias (2nd and 3rd column) are conditional upon rain: the "Wet-Gauge" Bias considers only hours where the gauge amounts are larger than 0.4 mm/h; finally, the definition of "wet-wet" Bias reduces further the number of hours used in the calculus: in such case, only hours with both radar and gauge amounts larger than 0.4 mm/h are used.

Site	"Overall" Bias	"Wet-Gauge" Bias	"Wet-Wet" Bias
Palermo radar	−0.9 dB (144 h)	−1.1 dB (48 h)	−0.7 dB (38 h)
Bisacquino radar	−1.0 dB (96 h)	−1.2 dB (37 h)	−0.9 dB (29 h)
Torino radar	−0.3 dB (126 h)	−0.4 dB (63 h)	+0.5 dB (41 h)

Table 5. QPE evaluation summary in terms of Bias for three mini-radar sites. The number of hours of each data set are given in parentheses.

Table 6 summarizes the dispersion of the radar-gauge errors using the hydrological-oriented score called **Scatter** (see Sec. 4.1.2). Since, as discussed previously, in the radar detection process (see for instance the "multiplicative" nature of the meteorological radar equation derived by *Probert-Jones* [1962]), the multiplicative nature of error prevails, the Scatter is defined as a ratio between the Radar (the device under test) and the Gauge (the reference). Hence, dry hours cannot be considered in evaluating the Scatter (unless using some trick, like for instance adding a negligible amount...) Consequently, only "Wet-Gauge" hours or "wet-wet" hours are considered in Table 6.

It can be concluded that the three-presented X-band radars are less reliable at low rain rates. By limiting the observations to hours with both Radar and Gauge amounts larger than 0.4 mm/h, the agreement improves not only in terms of Bias, but most of all in terms of Scatter. Finally, in the interpretation of these values of Bias and Scatter, it is important to bear in mind the large intrinsically different sampling modes as well as mismatches in time of the radar and gauge devices (e.g. *Zawadzki* [1975]).

Site	"Wet-Gauge" **Scatter**	"Wet-Wet" **Scatter**
Palermo radar	**2.38 dB** (48 h)	**1.97 dB** (38 h)
Bisacquino radar	**1.47 dB** (37 h)	**1.31 dB** (29 h)
Torino radar	**5.38 dB** (63 h)	**3.73 dB** (41 h)

Table 6. QPE evaluation summary in terms of Scatter for three mini-radar sites. The number of hours of each data set are given in parentheses.

5. Open issues and limitations

Short-wavelength (X-band) radar has the benefit of attaining high spatial resolution with a smaller antenna. However, there is a clear disadvantage compared to longer wavelengths: an increased attenuation in the presence of precipitation. Imagine three 2-km convective cells with instantaneous rain rate of 20, 40 and 100 mm/h respectively: at X-band frequencies these cells would cause two-way attenuation in radar reflectivity values of approximately 1.5, 3.6! and 11!! dB. Such figures preclude not only the use of X-band radar for long-range monitoring but also an accurate QPE, even at short-range. A partial remedy could be the use of polarimetric information, but this would remarkably increase the cost of the system: in the interesting work by *Mc Laughlin et. al* [2009], the cost of a Doppler, fully Polarimetric advanced X-band system developed within the framework of the CASA project is estimated in approximately 180 kEuro, which is almost 8 dB more expensive than our low-cost, semi-quantitative approach.

6. Summary: Filling the gap, which is observing the lowest part of the troposphere at short-range with portable, low-cost radars

Radar sampling volume increases with the square of the range (beam broadening) therefore, at longer ranges, small but intense features of the precipitation system are blurred (non-homogeneous beam filling). Furthermore, it is more likely to include different types of hydrometeors (e.g. snow, ice, rain drops), especially in the vertical dimension. At long-range, because of the decreased vertical resolution, the lower part of the sampling volume can be in rain whereas the upper part can be even characterized by an echo weaker than the radar sensitivity itself (beam overshooting).

By the term "range degradation" we mean several important sources of uncertainty regarding radar-based estimates of rainfall: beam broadening, non-homogeneous beam filling, partial beam occultation, overshooting and, depending on the operating frequency, attenuation. Such sources of uncertainty in general increase with increasing range. Current approaches to operational weather observation are based on the use of physically large, high-power, long-range radars, which are blocked from viewing the lower part of the troposphere by the Earth's curvature combined with orography. Hence, range degradation is one of the main problems in QPE and certainly a key factor in the underestimation of rainfall accumulation at far ranges with conventional long-range radars (e.g. *Kitchen and Jackson* [1993], *Smith et al.* [1996], *Meischner et al.* [1997], *Seo et al.* [2000], *Gabella et al.* [2000], *Chumechean et al.* [2004], *Joss et al.* [2006]).

This Chapter describes an alternate approach based on networks of large number of small, low-cost, X-band radars. Spacing these radars twenty kilometers apart defeats the Earth's curvature problem and enables the sampling of the lowest part of the troposphere using small antennas and low-power transmitters. Such networks can provide observing capabilities which supplement the operational state of the art radar network satisfying at the same time the needs of multiple users. Improved capabilities associated with this technology include low-altitude coverage and high temporal resolution. This technology has the potential to supplement the widely spaced networks of physically large high-power radars in use today.

Indeed, short-wavelength low-cost radar is able to fill a remarkable gap in observational meteorology: small, low-cost, radars can be used to supplement conventional, long-range radar networks in complex orography regions (e.g. the Alps and the Apennines), in highly populated areas (improving urban hydrology in major towns), in sensitive regions (areas prone to hydro-geological hazards) and along technological networks (e.g. highways, gas pipelines, ...) New important spatio-temporal scales (see Table 7), which characterize the highly variable precipitation field can now be investigated at affordable costs thanks to portable, low-cost, X-band weather radar developed, among others, by the Remote Sensing Group at Politecnico di Torino.

Type of device	Band	Cost	Coverage	Sampling vol.	Temporal res.
TRMM (or future GPM) spaceborne radar	Ku (Ka)		global	$5 \cdot 10^9 - 10^{10}$ m^3	Once per day
Long-range, Doppler, dual-pol radar	S (C)	1000 to 2000 k€	200 000 km^2	$10^5 - 10^9$ m^3	300 s
Medium-range dual-pol radar	X	200 to 500 k€	5000 km^2	$10^4 - 10^8$ m^3	120 s
Short–range radar	X	30 k€	2000 km^2	$10^4 - 10^7$ m^3	30 s
Rain Gauge	---		point	50 m^3	600 s
Disdrometer	---		point	10 m^3	1800 s

Table 7. The observational gap filled by the new portable, low-cost X-band weather radar network.

7. Acknowledgements

Gauge data were provided by the Regione Autonoma Valle d'Aosta – Ufficio Meteorologico della Protezione Civile, the Servizio Informativo Agrometeorologico Siciliano and Weather Underground. The mini radar development was possible thanks to the financial and technical support by Consorzio Interuniversitario per le Fisiche della Atmosfera (CINFAI).

8. References

Chumchean, S., Seed, A., and A. Sharma, 2004: Application of scaling in radar reflectivity for correcting range dependent bias in climatological radar rainfall estimates, J. Atmos. Ocean. Technol., 21, 1545-1556.

Doelling, I. G., J. Joss, and J. Riedl, Systematic variations of Z-R relationships from drop size distributions measured in Northern Germany during seven years, Atmos. Res., 48, 635-649, 1998.

Gabella, M., M. Bolliger, U. Germann, and G. Perona, 2005: Large sample evaluation of cumulative rainfall amounts in the Alps using a network of three radars, Atmos. Res., 77, 256-268.

Gabella, M., J. Joss, and G. Perona, 2000: Optimizing quantitative precipitation estimates using a non-coherent and a coherent radar operating on the same area, J. Geophys. Res., 105, 2237-2245.

Gabella M., Joss J., Michaelides S., Perona G., 2006: Range adjustment for Ground-based Radar, derived with the spaceborne TRMM Precipitation Radar, IEEE Trans. on Geosci. Rem. Sens., 44, 126-133, doi: 10.1109/ TGRS.2005.858436

Gabella M., Orione F., Zambotto M., Turso S., 2008: A Portable Low-Cost X-Band Radar For Rainfall Estimation In Alpine Valleys – Analysis of radar reflectivities and comparison between remotely sensed and in situ measurements, FORALPS Final Meeting Report, ISBN 978-88-8443-235-3, pp. 39.

Gabella M., Morin E., Notarpietro R., 2011a: Using TRMM spaceborne radar as a reference for compensating ground-based radar range degradation: Methodology verification based on rain gauges in Israel, J. Geophys. Res., 116, D02114, doi: 10.1029/2010JD014496

Gabella M., Morin E., Notarpietro R., Michaelides S., 2011b: Precipitation field in the Southeastern Mediterranean area as seen by the Ku-band spaceborne weather radar and two C-band ground-based radars, Atmos. Res., doi: 10.1016/j.atmosres2011.06.001.

Germann U., Galli G., Boscacci M., Bolliger M., Gabella M., 2004: Quantitative precipitation estimation in the Alps: where do we stand?, Third European Conference on Radar meteorology ERAD2004, Visby, Sweden, 2-6.

Hogan, R. J., O'Connor E. J. and A. J. Illingworth 2009: Verification of cloud-fraction forecasts, Q. J. Royal Meteorol. Soc., 135, 1494-1511.

Joss J., Gabella M., Michaelides S., and G. Perona, 2006: Variation of weather radar sensitivity at ground level and from space: case studies and possible causes, Meteorol. Z., 15, 485-496, doi: 10.1127/0941-2948/2006/0150.

Kitchen, M., and P.M. Jackson, 1993: Weather radar performance at long range simulated and observed, J. Appl. Meteor., 32, 975-985.

Marshall, J. S. and W. M. Palmer, 1948: The distribution of raindrops with size, J. Meteor., 5, 165-166.

Mc Laughin and other 28 coauthors, 2009: Short-wavelength technology and the potential for distributed networks of small radar systems, Bull. Am. Meteorol. Soc., 90, 1797-1817.

Meischner, P., Collier, C., Illingworth, A., Joss J., and W. Randeu, 1997: Advanced weather radar systems in Europe: the COST 75 action, Bull. Am. Meteorol. Soc., 78, 1411-1430.

Murphy, A. H., 1996: The Finley affair: A signal event in the history of forecast verification,Weather Forecast., 11, 3-20.

Notarpietro R., Zambotto M., Gabella M., Turso S., Perona G., 2005: The radar-ombrometer: a portable, low-cost, short-range X-band radar for rain estimation within valleys, VOLTAIRE final conference joint with the 7th European Conference on Applications of Meteorology (ECAM7) and the European Meteorological Society meeting (EMS05), September 12-16, Utrecht, The Netherlands, 19.

Probert-Jones, J. R., 1962: The radar equation in meteorology, Quart. J. Royal Meteorol. Soc., 88, 485-495.

Seo, D. J., Breidenbach J., Fulton R., Miller D., and T. O'Bannon, 2000: Real-time adjustment of range-dependent biases in WSR-88D rainfall estimates due to non-uniform vertical profile of reflectivity. J. Hydrometeorol, 1, 222-240.

Smith, J. A., Seo D. J., Baeck M. L., Hudlow M. D., 1996: An inter-comparison study of NEXRAD precipitation estimates, Water Resour. Res., 32, 2035-2045.

Tartaglione N., 2010: Relationship between precipitation forecast errors and skill scores of dichotomous forecasts, Weath. Forec., 25, 355-364.

Young C. B., Nelson B. R., Bradley A. A., Smith J.A., Peters-Lidard C. D., Kruger A., Baeck M. L., 1999: An evaluation of NEXRAD precipitation estimates in complex terrain, J. Geophys. Res. Atmos., 104, 19691-19703.

Sauvageot, H., 1992: Radar Meteorology, Boston, Artech House.

Zawadzki I., 1975: On radar-raingage comparison, J. Appl. Meteorol., 14, 1430-1436.

Permissions

The contributors of this book come from diverse backgrounds, making this book a truly international effort. This book will bring forth new frontiers with its revolutionizing research information and detailed analysis of the nascent developments around the world.

We would like to thank Joan Bech and Jorge Chau, for lending their expertise to make the book truly unique. They have played a crucial role in the development of this book. Without their invaluable contribution this book wouldn't have been possible. They have made vital efforts to compile up to date information on the varied aspects of this subject to make this book a valuable addition to the collection of many professionals and students.

This book was conceptualized with the vision of imparting up-to-date information and advanced data in this field. To ensure the same, a matchless editorial board was set up. Every individual on the board went through rigorous rounds of assessment to prove their worth. After which they invested a large part of their time researching and compiling the most relevant data for our readers. Conferences and sessions were held from time to time between the editorial board and the contributing authors to present the data in the most comprehensible form. The editorial team has worked tirelessly to provide valuable and valid information to help people across the globe.

Every chapter published in this book has been scrutinized by our experts. Their significance has been extensively debated. The topics covered herein carry significant findings which will fuel the growth of the discipline. They may even be implemented as practical applications or may be referred to as a beginning point for another development. Chapters in this book were first published by InTech; hereby published with permission under the Creative Commons Attribution License or equivalent.

The editorial board has been involved in producing this book since its inception. They have spent rigorous hours researching and exploring the diverse topics which have resulted in the successful publishing of this book. They have passed on their knowledge of decades through this book. To expedite this challenging task, the publisher supported the team at every step. A small team of assistant editors was also appointed to further simplify the editing procedure and attain best results for the readers.

Our editorial team has been hand-picked from every corner of the world. Their multi-ethnicity adds dynamic inputs to the discussions which result in innovative outcomes. These outcomes are then further discussed with the researchers and contributors who give their valuable feedback and opinion regarding the same. The feedback is then collaborated with the researches and they are edited in a comprehensive manner to aid the understanding of the subject.

Apart from the editorial board, the designing team has also invested a significant amount of their time in understanding the subject and creating the most relevant covers. They scrutinized every image to scout for the most suitable representation of the subject and create an appropriate cover for the book.

The publishing team has been involved in this book since its early stages. They were actively engaged in every process, be it collecting the data, connecting with the contributors or procuring relevant information. The team has been an ardent support to the editorial, designing and production team. Their endless efforts to recruit the best for this project, has resulted in the accomplishment of this book. They are a veteran in the field of academics and their pool of knowledge is as vast as their experience in printing. Their expertise and guidance has proved useful at every step. Their uncompromising quality standards have made this book an exceptional effort. Their encouragement from time to time has been an inspiration for everyone.

The publisher and the editorial board hope that this book will prove to be a valuable piece of knowledge for researchers, students, practitioners and scholars across the globe.

List of Contributors

Dusan S. Zrnic
NOAA, National Severe Storms Laboratory, USA

Paul Joe
Environment Canada, Canada

Sandy Dance
Bureau of Meteorology, Australia

Valliappa Lakshmanan
CIMMS/OU/National Severe Storms Laboratory, USA

Dirk Heizenreder, Paul James, Peter Lang and Thomas Hengstebeck
Deutcher Wetterdienst, Germany

Yerong Feng
Guandong Meteorological Bureau, China Meteorological Agency, China

P.W. Li and Hon-Yin Yeung
Hong Kong Observatory, China

Osamu Suzuki and Keiji Doi
Japan Meteorological Agency, Japan

Jianhua Dai
Shanghai Meteorological Bureau, China Meteorological Agency, China

P.W. Chan
Hong Kong Observatory, Hong Kong, China

Pengfei Zhang
University of Oklahoma, Norman, OK, USA

Clive Pierce
Hydro-Meteorological Research, Met Office, Fitz Roy Road, Exeter, UK

Alan Seed
Centre for Australian Weather and Climate Research, Bureau of Meteorology, Melbourne, Australia

Sue Ballard, David Simonin and Zhihong Li
Joint Centre for Mesoscale Meteorology, Met Office, Meteorology Building, University of Reading, Earley Gate, Reading, UK

Elena Saltikoff
Finnish Meteorological Institute, Finland

Ernesto Caetano and Víctor Magaña
Instituto de Geografía, National Autonomous University of Mexico, Mexico

Baldemar Méndez-Antonio
Energy Department, Metropolitan Autonomous University, Mexico

Marco Gabella
Meteoswiss, Switzerland
Politecnico di Torino – Electronics Department, Italy

Riccardo Notarpietro and Andrea Prato
Politecnico di Torino – Electronics Department, Italy

Silvano Bertoldo, Claudio Lucianaz, Oscar Rorato, Marco Allegretti and Giovanni Perona
Consorzio Interuniversitario per la Fisica delle Atmosfere (CINFAI) – Sede di Torino, Italy